D0999891

Bricks
to build a house

Also by John Woodforde

The Strange Story of False Teeth
The Truth about Cottages
The Story of the Bicycle
The Strange Story of False Hair

Bricks

to build a house

by John Woodforde

Routledge & Kegan Paul

First published in 1976
by Routledge & Kegan Paul Ltd
Broadway House, 68–74 Carter Lane,
London EC4V 5EL
Set in 10pt Ionic
Filmset and printed in Great Britain by
BAS Printers Limited, Wallop, Hampshire

ISBN 0 7100 8105 7

A London Brick
Publication

Contents

Illustrations

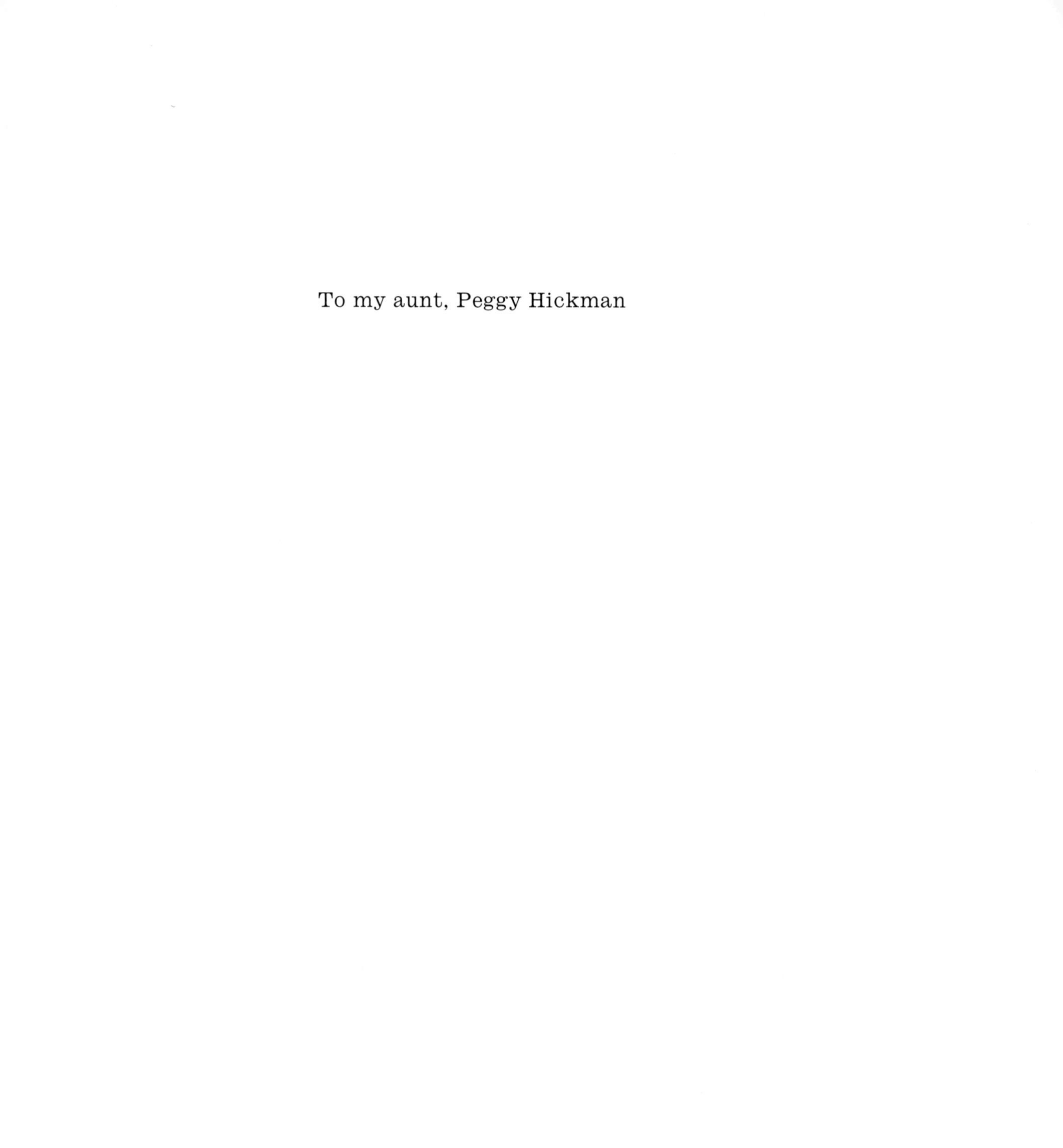

To my aunt, Peggy Hickman

Acknowledgments

I should like to thank many people for their help, particularly Mr Jeremy Rowe for suggesting this book and for his subsequent encouragement; Sir Ronald Stewart, Bt, chairman of London Brick Company Limited, for access to records of the fletton industry; Mr J.P. Bristow for allowing me to make use of his notes and for correcting a typescript; Mr Morris S. Whitehouse and Mr R. Richards for reading a typescript and much advice; Mr Kim Leslie for details about a private brickfield; Mr Basil Butterworth, Mr H.W.H. West and Mr Roy Edwards for technical information; the late Mr Geoffrey Laurence for original thoughts about brick and for allowing me to select from his collection of colour transparencies; and Mrs Ronald Sparks for careful re-typing.

To several members of the British Brick Society—which exists to promote interest in the history of brick—I am grateful for a wealth of additional facts and amendments generously supplied, though nothing in the book represents the official judgment of this society.

J.W.

1 Burnt earth

Several English churches near brickfields include in their harvest display a heap of unfired bricks.* Since the material for these comes from just below the surface of the earth, it is not entirely fanciful for clergymen to regard 'green' bricks as fruits of the soil needing, like flour, only cooking to make them perfect. Indeed, bricks and bread, said to be Britain's most basic industries, were in 1973 considered together by the Monopolies Commission.

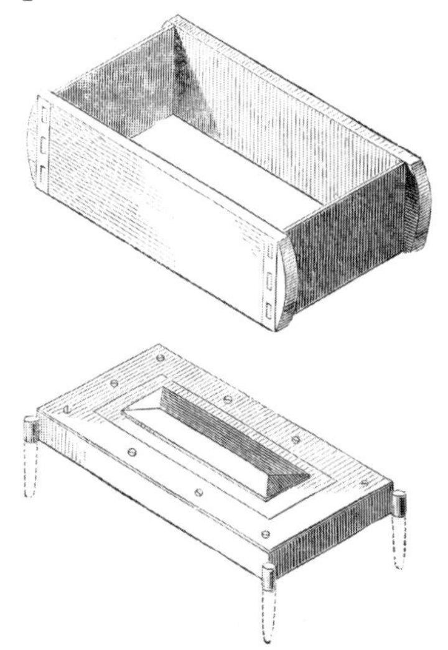

1
Mould (top) and stock-board (below) of the kind used for making London's bricks in the nineteenth century.
At each corner of the stock an iron pin was driven into the moulding table; the thickness of the brick could be regulated by the distance to which the pins were driven down

Brickfields were formerly a concern of missionary clergymen because of the degenerate behaviour that went on in them; few occupations were thought meaner than brickmaking and few kinds of workmen rougher than those who followed it.

* At Biddenham, Bedfordshire, fired ones are annually displayed in the church—on Rogation Sunday.

2
Some of the workers were very young—print of 1871

Some of the workers were very young. The *Graphic* of June 1871 wrote of drunkenness in brickyard children under ten and reported that at Oldbury, Lancashire, 'a child, four years old, was found helping her sister aged seven to carry clay.'

Charles Dickens in *Bleak House* (1853) makes his most brutish character a brickmaker, and describes in some detail the interior of his cottage, his wife's bruises and his daughter's laundry arrangements. Anthony Trollope in *Framley Parsonage* (1867) writes of a West Country parish 'abounding in brick-makers, a race of men very troublesome to a zealous parson who won't let men go rollicking to the devil without interference'. But however troublesome and rollicking, the brickmakers of the past matched the hard-drinking railway navvies in producing results which endured and are, indeed, an important part of Britain's heritage.

The British, living in a land of brickwork, have long seen the brick as a symbol of solidity. We talk of an object being as hard as a brick, of knocking heads against brick walls and of sinking like a brick. We drop a brick. It is probably a hard brick, likely to break something; but in this metaphor there can be a reminder of the softness of a brick when just moulded and of the need for careful handling. To pay someone the compliment of calling him a brick, though now an entirely English expression, is believed to derive from the reply of an Eastern king to an invading army which made fun of his unwalled towns: 'My troops are my walls and every soldier is a brick.'

The brickmaker no longer symbolises the rough-living worker, but there are still brickyards in which he works in the old way. Electrically driven machines may dig out and mix the raw material, but some of the best bricks are moulded one at a time by hand. The moulders are paid by the piece and may turn out over 1,500 bricks in a day. To watch a strong young man raising a lump of clay above his head, dashing it generously into the mould, and then striking off the surplus, offers as picturesque a sight as any in modern manufacture.

Out in the country the places of work can be picturesque, too, even romantic, with their rustic drying and making sheds and primitive kilns in a setting of hollows, mounds and hillocks. In L.P. Hartley's celebrated novel *The Brickfield*, an isolated brickyard in the fens of Suffolk is the scene of innumerable secret meetings between a young boy and his girl. All over the world, it seems, a rural brickyard may offer itself as a halting place, an oasis of warmth, during a walk on a cold evening. The hero of a modern Chinese novel,* strolling with a girl, says he will take her to a nice place.

> 'Where are we going?' She followed him.
> 'There is a brickyard on the southern slope. It will be warm there.'
> When they got to the southern slope, they could see smoke rising from the brick-kiln chimney. Nearby was a thatched woodshed, facing south and backing on the kiln. They went into the shed and it was very warm. They sat side by side on a bundle of faggots. The moon beams slanted in from the West under the low straw eaves . . .

Proprietors of country brickyards in Britain generally put up a notice to the effect that visitors after hours come at their own risk. They don't mind couples but are worried about tramps who leave litter, or worse, to be found in the morning. Mr H.P. Pycroft, a Hampshire brickmaker and builder, can tell of a tramp who was found insensible from having gone to sleep hard against a clamp kiln at a point where fumes were escaping.

About 2 per cent of Britain's bricks are made by hand in small yards with buildings like old farm buildings, but these are the bricks which many people like to have in the parts of walls that show and high prices are paid for them. It is the hand-

* Chou Li-po, *Great Changes in a Mountain Village*. Extract given in John Gittings's *A Chinese View of China*, BBC, 1973.

3
Brickmaking explored as a subject for composition in W. H. Pyne's *Picturesque Groups for the Embellishment of Landscape*, 1845. This 'Encyclopaedia of Landscape opens a new field to the student of the picturesque', it was claimed. 'It may even be useful to the advanced artist.' Pyne was impressed by the speed of brick-making and by the way workers supplied one another 'without confusion or intermission'

made bricks, with their tell-tale marks, which the many mechanised brick factories seek to imitate. To compare such bricks with concrete products is to realise that they are the most natural-looking of the artificial building materials.

Although no machine-made brick can look exactly like the hand-made article, Britain's London Brick Company can now produce mechanically, and in vast quantities, facing bricks that agreeably suggest it: this is done before firing by coating fletton bricks with sand and a variety of other finishes.

The appeal of brick in general seems bound up with the simplicity of the basic process. To make an experimental bricklet no tool is necessary except perhaps a spade. Scoop out a lump of clayey subsoil from the garden, knead and shape it (as in breadmaking), slowly dry and then bake red hot.

Firing might seem a problem, but I have found that this is satisfactorily achieved by leaving the bricklet in the heart of a solid-fuel fire which is kept going for about thirty-six hours. The bricklet is more likely to turn out red than any other colour because of the prevalence of iron oxide in clay.

Brickmaking material is broken-down rock of one geological age or another and is conveniently called clay (or, by non-geologists, brickearth); but the amount of true clay present may be less than 10 per cent. True clay consists of alumino silicate minerals in particles of infinitesimal fineness and is distorted too much by heat to be useful on its own for brickmaking. It is only necessary that there should be enough of it mingled with the mass to give plasticity, the ability to hold a shape: the non-clay materials like sand serve the invaluable purpose of reducing shrinkage on drying and preventing the bricks from cracking under heat.

Some brickmakers are lucky enough to be able to dig material which naturally contains clay in the best proportion for commercial brickmaking. A mixture of clay and sand called loam is one of these; another is a mixture of clay and chalk called malm.

4
A photograph of 1861 showing brickmaking at Barking Creek, Essex

Less fortunate brickmakers must be always adjusting their clay to give it a suitable plasticity. Very often sand or finely ground ash is added. The blue London clay is a good example of a clay which is far too sticky to be used on its own without a heavy admixture of other materials to reduce plasticity. Some of the shales, which are hardly sticky at all, are made plastic by grinding to powder and adding water.

Water-soluble salts in the clay can cause efflorescence on the finished brick unless appropriate action is taken during the preparation and during kilning; and stones, especially lumps of limestone which swell on getting wet, can cause disintegration. One advantage of the old method of tempering clay for moulding with the bare feet was that stones could be readily located by feel and thrown out. Now they are generally dealt with by powerful crushing.

There are three main ways of making bricks, of which the so-called soft mud process is the most traditional. The clay mix is made into a paste by adding water, pushed into wooden moulds and at once tipped out: this operation is done by machine as well as by hand and the product is generally known as a stock brick, the term coming from the stock (or fixed piece of shaped wood) which holds the mould. Those who want stock bricks for their texture of minor irregularities and fold marks will probably not find it in the machine-moulded stocks.

In the wire-cut process, a fairly stiff mix is forced in a column through a rectangular die and cut off into bricks by taut wires. About a third of all British bricks are wire-cuts. Slightly under a half of the clay bricks are made by the semi-dry process of London Brick Company in which the clay, being hard, is first powdered and then pressed into shape.

To acquire their characteristic durability, bricks must be fired for a period of days at a temperature greater than 900°C, the actual temperature depending on the clay being used. As the heat rises the character of the clay and its colour continues to change. Where bricks of exceptional hardness are wanted—bricks of the sort called engineering—the clay is fired in such a way that vitrification (or partial fusion) of the

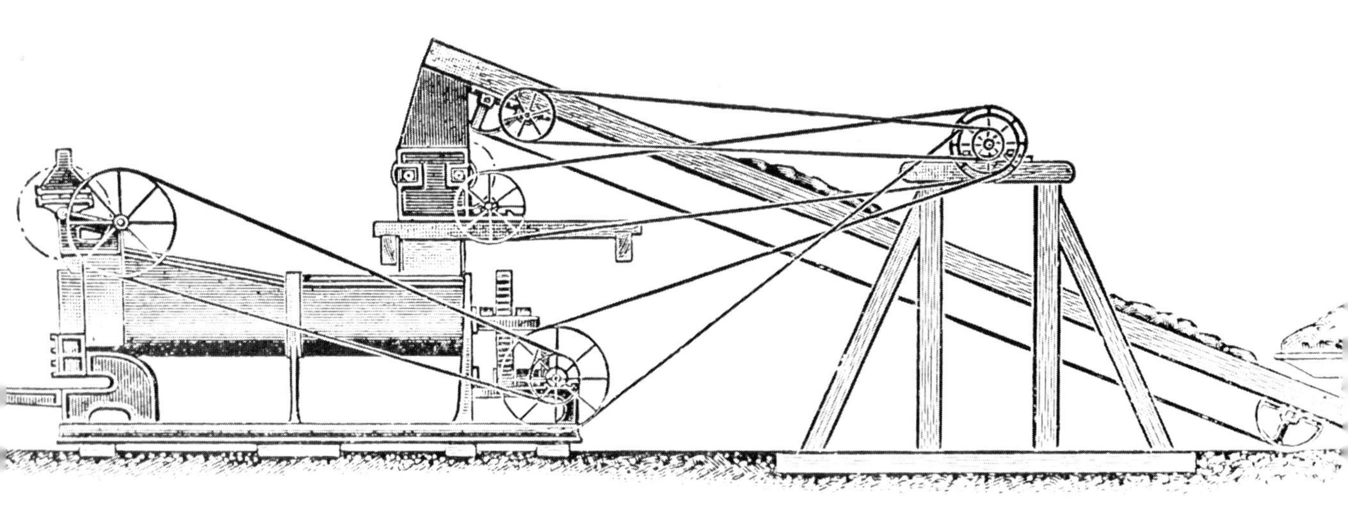

5
Simple plant for making machine-moulded bricks. On its way to the machine, the clay passes through a pair of crushing rollers to deal with small stones

material is pushed further than usual. The making of such bricks, and the forming in them of blue ferrous silicate, calls for careful firing because prolonged burning can easily lead to distortion brought about by the irregular behaviour of fused material. Both Staffordshire and Holland are noted for extra hard bricks. The small Dutch clinker, 7 inches by 3 by 1½, has in the past been imported for paving yards and stables.

Variations in clays, even within the same district, lead to corresponding variations in process from one brickyard to another; and in some there must be constant experiment in adding more or less of certain materials. The necessity for every commercial brickyard to adjust its way of brickmaking to the clay available seems to account for the fact that no industry has fewer handbooks on how to do it than brickmaking—less than a dozen between 1622 and 1966.

A truth about brickmaking which in recent years has taken many firms by surprise is that bricks emerge a slightly different colour when, for reasons of convenience, a kiln is fired by gas instead of coal. A large firm in the Midlands, supplying the bricks for a block of flats in Cheyne Walk, Chelsea, made the change in the course of meeting their order with the result that the bricks in the final delivery did not quite match the rest. The bricklayers, to the architect's dismay, continued to build with them. In the end, the huge expense of rebuilding was avoided by the procedure of the three royal gardeners in Carroll's *Alice*, the ones who planted white roses in mistake for red and were in danger, if the Queen found out, of having their heads cut off. The brick firm sent three men with paint and brushes and instructions to tint each brick in the upper storeys of the building to the shade of those in the lower storeys.

The properties of brick 2

Good brick is unlike concrete and stucco in needing no maintenance or surface treatment. It is improved by weathering and even looks the better for the passage of five hundred years. It is often more durable than natural stone. The London stock and yellow gault bricks with which Sir Joseph Bazalgette built the great sewers of London between 1858 and 1875 continue to serve their purpose. Concrete has been tried in recent times for sewer works in London, but bricks have been shown to wear better under the scouring of miscellaneous fluids and grit, to form stronger and more satisfactory junctions and to be less slippery for the sewer men to walk on.

6
The effect of a run-away lorry crashing into a house of 9-inch solid brickwork at Stoke on Trent—gratifying to brick men in that the damage is more confined than with other forms of construction

The handy weight and shape* makes brick relatively untiring to build with, for the workman can grasp his brick with one hand while picking up mortar with the other; he finds it, too, a flexible unit with which to follow most drawings. Even throughout the great stone building areas of the Mediterranean, it may be seen again and again in older houses that window and doorway arches, set like raised eyebrows in the stonework, are neatly and strongly brick. However, the extent to which builders have turned to bricks because of their regularity is traditionally hidden by a stucco finish—there are plenty of houses in the old quarters of Paris which contain less of the local coarse limestone than at first appears.

Having been hardened by intense heat, bricks are very good at resisting fire. Thus the metal in steel-framed buildings may be buried for safety in brickwork; and it can often be seen that the flues and chimney stacks of old houses, stone as well as wooden,

* The size of a brick has scarcely changed in centuries. The most recent British Standard is $8\frac{5}{8}$ inches by $4\frac{1}{8}$ by $2\frac{5}{8}$ or 215 mm by 102·5 by 65 (B.S. 3921).

BERRESFORD
582 YTD

are of brick. Few bricks lack toughness, but obviously some must be used with common sense. There are less than fully-fired bricks which should not be exposed, unrendered, in the outer leaf of a wall; the otherwise robust fletton bricks are damaged by frost if laid as a coping on parapets.

But in general where brickwork has failed in any respect, the fault is unlikely to be in the bricks themselves. Take the damp patch on a ground-floor wall of a country cottage. This could well be explained by such things as the absence of a damp course or the presence just outside of a banked-up flower bed. If chimney breasts upstairs are damp, the cause could lie in a lack of protective flashing at the junction of stack and roof.

It has never been a protection against rain to build with the hardest, densest bricks, for wherever water has encouragement to get in, it does so through chinks in the mortar joints rather than through the bricks themselves. In fact, bricks that are porous enough to absorb a certain amount of water actually discourage the entry of rain at the joints.

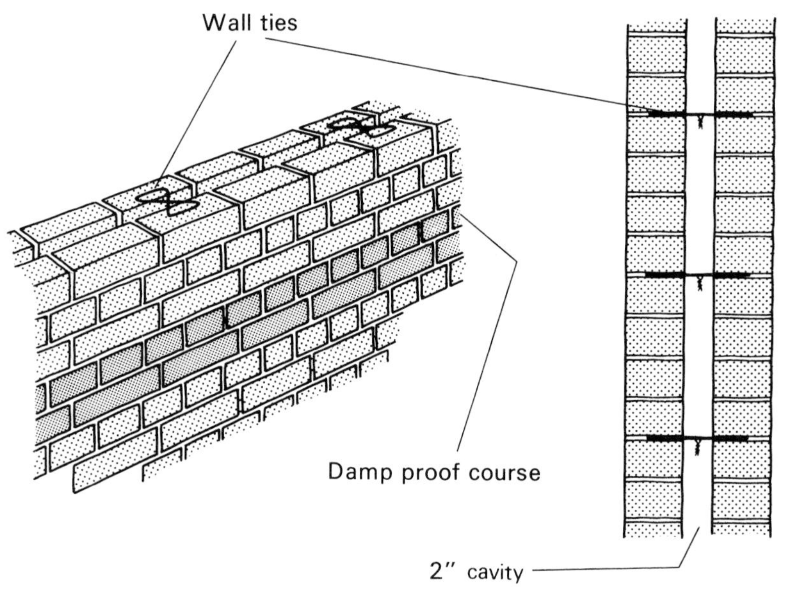

7
Cavity wall with metal ties in which the appearance of English bond is given by the use of half-bricks

The chances of penetration by rain are reduced to nil today by cavity-wall construction, which also gives a degree of insulation. And no water can rise up from the ground where a dampcourse is built in, as is usual, just below the level of the floor.

The insertion of dampcourses, formerly of slate, had become the usual practice by the end of the nineteenth century. On the other hand, the cavity wall, with metal ties to hold it together, was almost unknown until 1912. For years this construction was specified only occasionally for the more expensive brick houses. But to the busy speculative builders of the 1930s, it was welcomed, Morris S. Whitehouse remembers, 'as a boon and a blessing because it was almost foolproof against bad workmanship and bad design'. Unless a workman completely blocked up a cavity with mortar droppings, no water could get through.

It has been argued that the dampcourse was a more important invention than the cavity wall. The point is debatable. But it can certainly be said that while a house without a dampcourse is liable to be damp whether its walls are cavity or solid, a house without cavity walls need give no trouble if it is carefully built and if there are wide eaves and appropriate devices, especially below windows, to throw off water. The fact that many thousands of Britain's older houses remain dry with 9-inch solid brick walls is a tribute to sound workmanship.

All the same, quite serious failures in brickwork took place from time to time in the mid-nineteenth-century period because of skimped mortar and poor laying. Workers' houses in London which partially fell over in the course of construction made a not-uncommon item of news in the papers. During the building of Euston Station in 1848, inefficient cutting of bricks to the right shape for arches led to a disastrous collapse.

Old brick walls remain which, though still doing their job, are leaning over, cracked or bulged. The reason for such failures may become clear in the exposure brought about by demolition work. Tenuous foundations are found. Inside a misshapen solid brick wall it may be seen that the mortar is no stronger than dust, thus allowing the individual bricks far too much freedom to shift and settle. A twentieth-century cavity wall, perhaps not forty years old, yet having an ominous bulge, will suddenly expose its fragile-looking section and with it the fact that the ferrous metal wall ties have come adrift because of rust or settlement.

Brick buildings rarely develop a fault because the bricks are not hard enough. Yet since the nineteenth century the commonest laboratory test has been for their crushing (or compressive) strength. Even today the brick firms like to be conversant with these figures for their various grades of brick and no sales leaflet is thought complete without a statement of them.

Although customers may be persuaded that high crushing-strength figures indicate high quality, such figures need not be a reliable indication that certain bricks resist the weather well or are more serviceable than others. The figures below indicate the range of strengths in lb. per square inch with, in brackets, the equivalent in meganewtons or N mm^2:

Engineering bricks

Class A	10,000	(69)
Class B	7,000	(48·5)

Facing bricks

Flettons	3,000–4,500	(20·5–31·8)
Others	1,000–12,000	(7–83)

Common bricks

Flettons	3,000–4,500	(20·5–31·8)
London Stocks	750–1,500 or more	(5·2–10·5)
Others	2,000–6,000	(14–41·4)

Alfred Searle, in the 1956 edition of his *Modern Brickmaking*, points out that a brick in a wall has the tiny area of 0·25 square feet which is subject to vertical pressure from the bricks above it and that no less than 8,000 would be needed to crumble a brick in the lowest range of crushing strengths. These 8,000 bricks, one on top of the other, plus mortar, would reach a height of 1,750 feet.

The architect planning tall buildings of load-bearing brickwork, which must be economically costed, has of course to take account of crushing strengths. And there are structures for which the density and strength of blue engineering bricks are appropriate. But for ordinary small houses any modern building brick is much stronger than necessary for its job.

Even before firing, dried bricks may have a crushing strength in excess of 350 lb. to the square inch and could be used in their green state for building a house in Britain, if the walls were subsequently rendered against rain. A bungalow in the south of England was recently built with green bricks by accident—a most unusual accident—and several months passed before it was even noticed. Having been well dusted with orange sand, the bricks had an appearance that deceived both the lorry driver and the bricklayer to whom he delivered them. The buyer of the bungalow moved in, quite satisfied; but one day he noticed, while watering roses planted against his walls, that the brickwork was being washed away. Needless to say, the brick firm had his bungalow rebuilt with the fired article.

There exist numerous old engineering works—bridges, retaining walls for railways—which theoretically demanded the strongest bricks and did not get them. They were built with humble stocks and yet are still in good order after more than a hundred years. However, Victorian engineers calculated brickwork by rule of thumb and, tending to over-specify, got massive results—the early skyscrapers of New York were built in mass brickwork. London stock bricks have yielded almost monolithic structures, for these fairly weak bricks have hardened in London's smoky air through recrystallisation of the calcite, and at the same time the lime mortar between them has been hardened by carbon dioxide.

The history of building in brick appears to offer only one instance of disaster caused by too low a crushing strength—and that was largely a matter of builders' negligence in that a great weight was allowed to rest on a small area of brickwork. During the afternoon of 17 January 1878, an oyster-seller in the Haymarket, London, noticed that the doors of his tall narrow house on the corner of Panton Street stuck in their frames. Towards night, creaking sounds were heard and lumps of plaster fell from ceilings. The oyster-seller sent his assistants outside, but he himself lingered. He was killed outright when his house, together with the one next to it, fell over into the street.

In the enquiry that followed it was established that, as a result of alterations to the two houses to produce shop fronts, much of the weight of both had been supported by a cast iron pillar resting on the stump of an old party wall. The bricks below and around the base of this pillar were crushed to dust.

How far this London accident of 1878 was a factor in making people sensitive about the compressive strength of bricks is unknown, but the enquiry was fully reported in the *Builder* and it so happens that shortly afterwards activity in the testing of bricks (and stones) became widespread. Later that year, in Germany, work began on setting up certain extra regional testing laboratories which had been called for by the Union of German Architects' and Engineers' Societies, and during the 1880s conferences were held in Munich, Dresden, Berlin and Vienna to discuss the best ways of testing building materials. In France the École Nationale des Ponts et Chaussées began testing bricks in 1881.

The United States Army was required in 1884 to test the bricks of the New Pension House in Washington, and reported that they withstood a pressure of up to 10,000 lb. to the square inch—more than good granite would stand. The first serious investigation in Britain into brickwork strength was conducted by a committee of the Royal Institute of British Architects between

8 *right*
A sample from thirteen centuries. At the base, a first-century Roman brick. Next, a twelfth-century great brick

9 *above left*
First century BC.
Bricks coloured and glazed depicting a Persian archer – part of a frieze at the Palace of Darius at Susa

below
A store house at Tell Jemmeh, Western Negev Desert, c.450 BC

1895 and 1897; piers built with different kinds of brick, in mortars made with lime and with cement, were tested after three months and six months.

Research into bricks and the machines to make them has been going on throughout the twentieth century in the laboratories of the bigger brick firms—London Brick Company spends £100,000 a year on research—and in the national research stations to which the industry and government subscribe.

The British Ceramic Research Association has for years been subjecting samples of brickwork to artificial rain, frost and wind; and it owns a rig of 900 tons with which the strength of walls is tried out by applying pressure from the side as well as from above. Tests inspired by London's Ronan Point disaster of 1970, in which slabs high on a tower block came adrift when there was a gas explosion, have shown that properly bonded brickwork will withstand explosions better than more modern building components. It was also found that only in the most contrived situation was it possible to achieve a gas-explosion pressure approaching 5 lb. per square inch. The British Ceramic Research Association also runs a clay-testing service. Members may send a sample and learn—before investing money in plant—how far their raw material is suitable for bricks. Since a clay report can be done on a sample of a pound or two, the service is well used by potential brickmakers overseas. It is an achievement of the Association that one of its experts on bricks, Mr H.W.H. West, has written a useful book on brickmaking for the developing countries, published by the United Nations in 1969. It is read in parts of the world where the need is to employ as many people as possible, rather than as few, and it thus describes procedures and equipment which are now old-fashioned in Britain.

A lot of work has been done on methods of putting holes in bricks to make them lighter, to save clay and to keep out moisture. The Building Research Establishment at Watford was especially active over this during the 1950s and drew interest with the invention of a double brick called the V, designed to span the whole width of a wall—it consisted of two perforated bricks joined together with webs in such a way that penetration by water is almost impossible. It has not been taken up commercially, partly on the grounds of expense and partly because of some discomfort in handling.

Hollow clay blocks, or structural clay tiles as they call them in America, have always been strong enough, despite their brittle appearance, for external walls as well as load-bearing internal walls. But they could not be used in this way until 1937, when the Ministry of Health issued revised model by-laws for building. Previously the by-laws required that all external walls, at least in towns, should consist of not less than 8 inches of solid incombustible matter.

The change in the regulations did not cause a rush to make use of hollow clay blocks. Unlike in France, where they have been used in great quantities to replace stone since 1855 (solid bricks are today made only in a small area of the north of France), in Britain architects and builders, not to mention their clients, have never ceased to hold conservative views about them. But the prime reason for resorting only in a limited way to hollow blocks (for backing work and floors) is that traditional solid bricks are unusually cheap, especially the fletton kind which are economically fired with the fuel occurring naturally in the clay from which they are made. Low cost compensates for the rather longer time it takes to build a wall with bricks than with the larger hollow blocks.

3 Brick BC

The story of the brick begins with sun-dried mud bricks formed with hands alone. I have seen walls built with such bricks at least 10,000 years ago deep in excavated sections of ancient Jericho, which is believed to be the oldest town in the world. There are prized-out specimens clearly showing Neolithic hand marks.

These early bricks have the shape a person gives automatically to dough when breadmaking without a tin, and indeed they resemble long loaves of bread, flat underneath and rounded on top. Squeezing and consolidating, the makers sometimes left a bold pattern of thumb impressions; these had the incidental function of helping the bricks to key in with mud mortar.

The soil of Jericho, much of it windblown dust with a proportion of finely powdered lime, was especially suitable for sun-dried bricks. When turned to paste with water, it had the plasti-

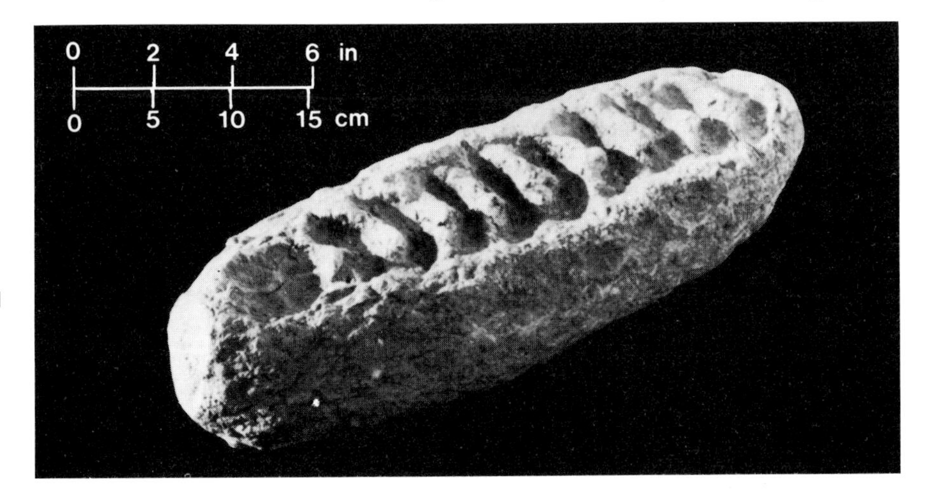

10
One of the oldest bricks in the world taken from a Pre-Pottery Neolithic house at Jericho—around 8000 BC. Of sun-dried mud, it was formed with the hands like bread and has thumb marks on the top. These helped to provide a keying with mud mortar

city of clay and became dense and heavy on drying. Neolithic bricks ring to the knock like concrete and have been shown by modern tests to have a crushing strength equal to half that of an ordinary fired brick. They were vulnerable to water, as are all plain mud bricks, and needed a plastering over with mud after wettings in the rainy season. The owner of a mud-brick house in modern Jericho assured me in 1974 that maintenance was no great trouble because it seldom rained.

Sun-dried bricks, requiring only labour to make, were so effective that kiln-fired bricks did not appear till the third millennium BC, long after the art of pottery had demonstrated the effect of a high temperature on clay. For thousands of years fired bricks were a luxury article for those able to command the use of scarce fuel. Even today, according to a recent United Nations survey, more of the world's houses are built of unfired earth than of any other material.

In several countries, including Russia, over half the houses are of this type. Mostly they are simple one-storey dwellings, but it is possible by thickening walls to build tall houses with mud bricks. In Anatolia, Turkey, where the manufacture of mud bricks is a centralised industry, houses built with them may be five storeys high; wide roof eaves, giving Turkish mud houses a Regency look, protect against the worst effect of rain, and bottom courses of stone prevent erosion by dampness. Mud and straw is moulded into outsize blocks in Mexico, where the word adobe is merely Spanish for mud: adobe has been a South American building material for at least 5,000 years.

The earliest peoples on earth were nomads who sheltered in caves and temporary huts. Making bricks came with settling down in one place and raising food from the land. Jericho drew the world's first settlers less because it was an oasis with a permanent and copious supply of water from the Jordan than because, by a chance thought miraculous, wheat instead of grass began to grow there. For people settling down, something more robust than shelters of twigs and daub was wanted, and the local soil, kneaded into convenient shapes, made the obvious building material. Neat mud bricks were quicker to build with than lumps of stone that were irregular and in any case hard to find at Jericho.

When the site of Jericho was first occupied the ground was flat and on a level with the river. Today it is a ten-acre grassy mound rising to seventy feet, built up layer by layer of the remains of 9,000 years' worth of mud houses; some of them collapsed after being abandoned and others through destruction by earthquakes and invaders. Unlike the components of stone houses, the bricks of ruined mud houses are unsuitable for re-use: successive waves of settlers simply built again on top of smoothed-out remains.

The resultant mound is known in Arab countries as a tell. Jericho's tell has been of special interest to archaeologists because of the famous story in the Old Testament concerning the capture of it by the Israelites in the fourteenth century BC and the fall of the town walls. The tell has been excavated several times in the past hundred years, but never more ably and scientifically than by teams directed by Dame Kathleen Kenyon in the 1950s.

Her conclusion that the stumps of mud-brick walls at the base of the tell are datable to the Pre-Pottery Neolithic period, to 8000 BC or even earlier, was based on Carbon 14 analysis of such organic material as wood and bones found in or near the walls. In her book *Digging Up Jericho* Dame Kathleen states that to form these walls—they are mainly 18 inches thick and some have brief stone footings—the majority of the bricks

> are laid as stretchers, three to the width of the wall, but occasionally there are groups of headers running the full width of the wall. The whole forms a structure which it is still quite hard work to demolish, and it is a very difficult task to extract a single intact brick.

Dimensions varied from house to house, but a common size was 10 inches by 3 wide and 3 deep, just small enough for the builder to hold with one hand; in this characteristic of being easily handled, valued by bricklayers today, early Neolithic bricks are more similar to modern bricks than the great slabs which came to be made with the aid of moulds in the third millennium BC—and which were usual in Roman times.

The archaeologists were able to study at Jericho two types of house belonging to the Pre-Pottery Neolithic Age. In the earliest period the houses are small and round, a translation into a solid structure of the nomad's flimsy hut. Then, around 5000 BC, following an abandonment of Jericho for several centuries, a more sophisticated type emerged, rectangular in plan. Some had sizeable rooms linked by wide openings. Almost certainly they were of one storey only and roofed with poles and reeds daubed thickly with mud: this kind of roof is still common in the Middle East and elsewhere.

Living-room floors in this type of house had a finish of lime plaster laid concave at the junction with walls, a device used today in hospitals to prevent dust from accumulating in awkward corners. The floors emerged so hard and shiny that the archaeologists found they could effectively swill them down with pails of water, imagining as they did so the same action being performed 7,000 years ago by a Neolithic housewife. The cooking in these later houses of around 5000 BC seems to have been done in a central courtyard; for layers of ash were found separated by thin layers of clay, the find suggesting that when the courtyard became offensively grimy a new surface was laid on top.

This second Neolithic-Age Jericho came to an abrupt end, it has been realised, around 4500 BC. But a period of desertion was not followed by progress in the art of living. Indeed the next occupation exhibits a retrogression. The new people knew how to make vessels of pottery—an advance on hollowing out pieces of stone—but in all other respects they were more primitive than their predecessors. There is evidence that some settlers, rather than build, lived in pits dug out of the ruins. Those who made bricks made them small and bun-shaped. In *Walls of Jericho* Lady Wheeler, who worked with Dame Kathleen, commented that they seemed 'idiotic':

> It is strange that such an inconveniently shaped brick should have caught the imagination of those far-off peoples: but evidently it did, because they were used by builders farther away to the east in the Tigris and Euphrates valleys at the end of the fourth millennium . . . it is highly improbable that the idea for such an idiotic brick should have occurred spontaneously to two different peoples.

The smallness of bun bricks would have meant less failures through cracking in the hot sun, but most likely their appeal was simply that given two hands full of mud, the shape was the least trouble to form. Children could make them, and very likely did so.

The step forward in brickmaking of using a box mould instead of merely the hands is dated to the Early Bronze Age, a period starting in the Middle East around 3000 BC. The celebrated walls of Jericho, begun at the beginning of this period, were largely built of box-moulded mud bricks. These are quite different from the hand-formed type in being large even slabs, often 14 inches by 10 in plan and 2 inches thick; their regularity made it possible to have thinner mortar joints. There are signs that animals' dung was sometimes mixed into Jericho's bricks as a strengthener, but the incorporation of such fibrous binding material as straw, which was to become a usual practice in Egypt, was never thought necessary because of the cohesive properties of the soil.

The earliest walls of Jericho were about 3 feet 6 inches thick at the base, but they were rebuilt and strengthened over so many centuries that the thickness became enormously increased.

To hold the massive leafs together, here and there huge wooden ties were inserted.

The worst threat to Jericho's defences came always from the earthquakes which take place roughly four times a century in the Jordan Valley. At intervals in the length of certain sections of the wall there are cavities about 3 feet square which seem to have been an anti-earthquake device. They localised a collapse in the manner of today's expansion joints: in places a section is seen to have fallen while the adjacent section, beyond one of the cavities, still stands about 10 feet high.

Attacks on the walls by enemies occasionally took the form of lighting fires against them which, so far from causing a breach, served only to strengthen and harden them. At the foot of one stretch was found a layer of ash 3 feet deep, derived from a huge pile of brushwood. The mass of bricks, 17 feet thick at that point, had been burned red all through, a process helped (compare clamp firing at a brickyard) by interior fuel in the form of wooden ties. Building material was never deliberately fired, so far as is known, in the Early Bronze Age; though here, for anyone who had cared to take notice, was a demonstration of the effect of prolonged and intense heat on earth.

11 *left*
Mud bricks in the walls of Jericho. At the bottom of the excavation can be seen the collapsed face of an early town wall of the Early Bronze Age with tumbled bricks above. On the left, the face of a subsequent town wall

12 *centre*
Remains of Iron Age houses at Jericho

13 *right*
In the foreground, the tell of ancient Jericho. Beyond it, a refugee town built with mud bricks by its occupants in the 1950s. The houses are now almost uninhabited and their gradual and not unsightly collapse under the impact of rain, dust storms and sun demonstrates how a tell is built up

Largely because of erosion, remains of Late Bronze Age settlement have almost entirely disappeared, and unfortunately no trace at all remains of the walls of Jericho which fell down dramatically when menaced by the children of Israel in the fourteenth century; archaeology can add only conjecture to the vivid description in the Old Testament's Book of Joshua. 'One can visualise', writes Dame Kathleen Kenyon,

> the Children of Israel marching round the eight acres of the town and striking terror into the heart of the inhabitants, until all will to fight deserted them when on the seventh day the blast of the trumpets smote their ears. But as to what caused the walls to fall down flat, we have no factual evidence. We can guess that it was an earthquake, which the excavations have shown to have destroyed a number of the earlier walls, but this is only conjecture. It would have been very natural for the Israelites to have regarded such a visitation as divine intervention on their behalf, as indeed it can be regarded.

Evidence of very early use of mud bricks and blocks has been found in many other parts of the Middle East, in Central and

South America, in Turkey and in India. Charles Steen, of the USA National Parks, has written:

> Adobes were extensively used in the Chicama Valley of Peru as early as 3000 BC . . . the bricks were cast in moulds which were made of crushed and flattened bamboo. The peoples of the desert kingdoms of South America plastered their adobe structures—pyramids, temples, houses—with either a mud plaster or a thin lime wash.

Since nothing has been discovered to link the ancient cultures of America with the Old World, it seems that brickmaking, like pottery, has been a lesson twice learned by man.

During the 1960s archaeologists working at Catal Hüyük, southern Turkey, found numerous remains of houses built before 5700 BC with slabs so heavy that two men were needed to lift them—these finds have been discussed in a book on Catal Hüyük by James Mellart (1967). The same kind of large slab went to build most of eighteen successive levels of mud walling, dating from the fourth millennium, which has been found at Eridu, Iraq.

The period 2800–2300 BC produced in Iraq its own characteristic type of moulded bricks. They were flat bottomed with a convex top, about 9 inches by 6 and about 2½ inches thick at the centre—the result of heaping mud into a mould and rounding it off with the hands. The bricks were commonly laid herringbone-fashion: this too appeared to be characteristic of the place and period. Two or three courses of bricks on edge leaning in one direction were followed by two courses of bricks laid flat, and then by two or three in which they sloped diagonally in the opposite direction.

From about 2000 BC till the fall of Babylon in the seventh century BC the normal mud bricks for important works resembled blocks of stone, 12 inches by 6 and 6 inches thick, or 14 inches square and only 4 inches thick. Some enormous structures were put up, with walls up to 20 feet in width. A tower at Erech had its outer surface protected by thousands of pieces of pottery hammered into the mud while it was still slightly plastic; later on it became usual to encase certain buildings at least partially with burnt bricks.

The firing of bricks has been carried out since the third millennium BC. Several specimens of this date from Indus Valley sites are in the Museum of Ancient Brick, Johnson City, Tennessee. Any people who had acquired the art of making pottery obviously knew how to do it—and how to make other simple clay products like drain pipes and troughs—but, for ordinary building purposes, there was reluctance to expend scarce wood fuel on making building materials. Sundried bricks, dense and heavy, were certainly more satisfactory than those burnt bricks which for economy reasons had been fired at too low a temperature.

Egypt has always lacked wood and her early mud brick industry, demanding plenty of straw, is widely known about because of references to it in the Old Testament. The Egyptians put chopped straw in the mix because the alluvial mud of the Nile was strongly plastic and needed fibres to stop it cracking. Their great slab-like bricks, weighing about 35 lb., were employed for many of the early pyramids, dating from around 2000 BC.

The Israelites during their bondage were obliged to make enormous quantities of these heavy bricks for the erection of such cities as Pithom and Rameses. Biblical accounts of their hardships are illustrated by tomb pictures which show them staggering under the weight of water pots, mounds of mud and yoke-loads of finished bricks. Taskmasters, whip in hand, are plentifully represented. Nevertheless the Israelite captives seem to have been far from conscientious at times. The late Sir William Willocks reported in *From the Garden of Eden to the Crossing of the Jordan* that he had

> picked out of old ruins in the Delta scores of bricks which contained nothing but straw daubed round with mud. These

> had undoubtedly been made by captives who were contemplating revolt. The taskmasters had furnished a sufficiency of straw for a certain tale of bricks, but the captives had hurriedly wasted it and delivered a totally inadequate number of bricks. . . . Those who acted in this way had begun to feel that they were not utterly helpless.

Deliberate wasting of straw accounts for the Pharaoh's renowned instruction to the taskmasters of around 1490 BC:

> Ye shall no more give the people straw to make brick. Let them go and gather it for themselves, and the tale of the bricks, which they did make heretofore, ye shall lay upon them.

The Pharaoh, Rameses II, was faced with industrial action.

But this is not the first recorded dispute in the building trade, for around 2250 BC thousands of brickmakers and bricklayers went on strike at the site of the Tower of Babel. Here was to have been a tower of burned bricks so high it would reach

heaven, but when it had reached a certain height the workers in a mass left off what they were doing and scattered. They had become confused, it seems, by the number of different languages they were expected to understand and, according to chapter 11 of Genesis, the confusion was divinely inspired, signifying displeasure over the idea of raising a ladder to heaven. Shortly after the departure of the workers most of the tower fell down.

George Every has suggested in *Christian Mythology*, 1970, that the Genesis account serves as a parable on the subject of young people undertaking with enthusiasm tasks which are beyond their strength and powers of organisation. 'As soon as this is discovered the groups dissolve in mutual misunderstanding and recrimination.'

In later Jewish legend, Mr Every points out, the building of the Tower of Babel was the first occasion for the sacrifice of human beings for the sake of a structure. If a man fell to his death it did not matter, but if a brick fell, it might take a year to replace. Women helping to make the bricks were not allowed to stop

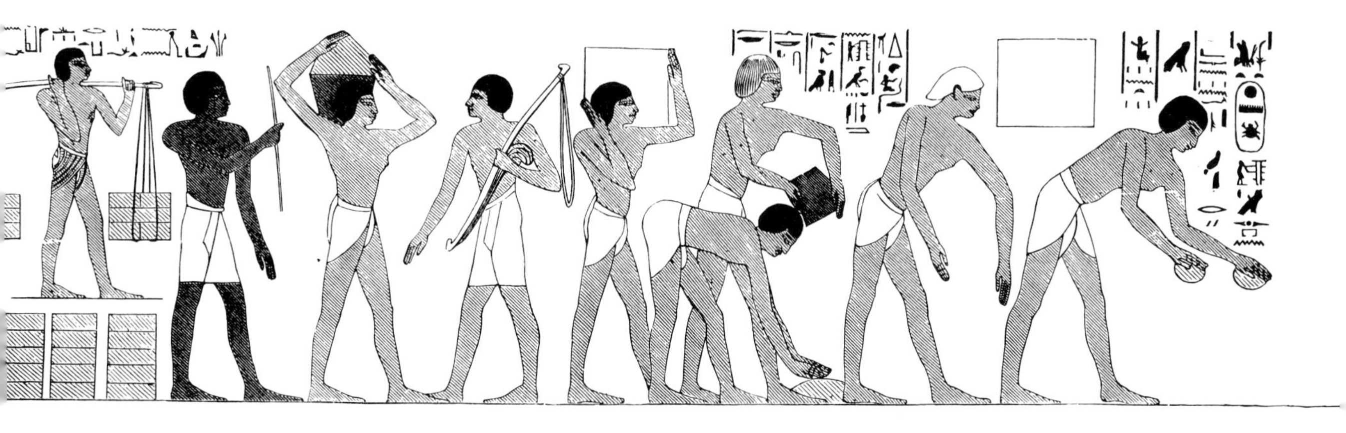

14
Egyptian wall paintings found in a tomb at Abd-el-Qurna which depict the making and use of bricks. They relate to the building of a temple dedicated to Amun. Middle Kingdom period

even to have a baby. When a baby was born, the mother had to strap it in a sheet to her back and go on working.

Where bricks were burnt four or five thousand years ago, the job was done by one of two methods which are still current—in England as well as in the Middle East. One method was to build a heap (or clamp) of green bricks, interspersed with fuel and flue tunnels, and set fire to it. The burning process would take several weeks, though less time and less fuel were needed with bricks containing straw, a built-in combustible matter. Dr Norman Davey suggests in his *History of Building Materials* that the average temperature reached would not have been much higher than 800°C, with the consequence that many clamp-fired bricks were underburnt and soft.

The other method was to burn the bricks in simple updraught kilns. These consisted of two chambers, one above for the bricks and one below for the wood fuel. An early example, dating from about 2500 BC, has been excavated at Khafaje. Similar kilns were built by the Romans, and kilns following the same principle are still in use in England at a few country brickyards.

By the end of the third millennium BC some elaborate work was being done with the aid of burnt brick in the Sumerian cities of present-day Iraq; there is evidence of a palace built at Erech (*c*. 2400 BC) and a temple at Ur (*c*. 2300 BC), while at Harappa and Mohenjo-Daro houses were built on platforms of burnt brick to raise them above flood level.

Most of the moulded bricks—that is, those not simply chopped into shape—were about 10 inches by 6 and 3 inches thick. The burnt bricks made centuries later by the Babylonians were commonly a little thicker and 12 to 18 inches square. By far the best example of building with these is the Ishtar Gate in Babylon constructed under Nebuchadnezzar II, 604–562 BC. It was faced with bricks embedded in bitumen; this was used hot, according to Herodotus, and came from Hit.

15
Kiln-fired bricks embedded in bitumen still face the buildings of the Ishtar Gate at Babylon built by Nebuchadnezzar ii (604–562BC); within the gate there are streets of brick set in bitumen. An inscription left by Nebuchadnezzar II states that the work in mud brick and mud mortar by his father did not resist the waters of the Euphrates. The fired bricks varied in size from 12 to 18 inches square and from 2 to $3\frac{1}{4}$ inches thick

Bituminous mixtures were also used in ancient Babylon for road-making—on a foundation that consisted of several courses of burnt bricks. After the collapse of the Babylonian Empire under Nebuchadnezzar the new rulers, the Persians and the Seleucids, gave up bitumen and used for their walls inferior loam or lime mortar. Well over two thousand years passed before bituminous mixtures were again used, for waterproofing buildings and for road-making. In 1835 the Place de la Concorde, Paris, was surfaced with bitumen mixed with powdered rock asphalt. In England the first road so treated was Threadneedle Street, London, in 1869. Nebuchadnezzar was 'modern', too, in introducing the making of glazed bricks for walls, some of them specially moulded to provide pictures of animals in relief.

Nebuchadnezzar's building work is given extra fascination by the fact that, unlike the Jerichoans, he lived in a civilisation which could write things down. In one of his inscriptions, translated by F. Delitzch in 1910, he describes his palace as the marvel of mankind, and explains that it had first been built by his father, King of Babylon, of *crude bricks*.

> In consequence of high waters, its foundations had become weak and because of the filling up of the streets the gateway of the palace had become too low. I tore down its walls of *dried bricks* and laid its cornerstone bare and reached the depth of the waters. Facing the water, I laid its stone firmly and raised it mountain high with bitumen and burnt bricks. Mighty cedars I caused to be laid down at length for its roofing. . . . For protection, I built two massive walls of asphalt and brick, 490 ells beyond Nimitte-Bel. Between them I erected a structure of bricks on which I built my kingly dwelling of asphalt and bricks.

The buried palace of Nebuchadnezzar has furnished material over the centuries for the building of innumerable houses. Bricks from this site have been noted in modern walls in Baghdad. Their identity is certain because of a custom, prevailing in Assyria as well as Babylonia, of marking each brick with the name of the king in whose reign it was made. Inscriptions in cuneiform characters were impressed on the brick in a sunken panel.

The end of a chapter mainly about the early mud brick, dried in the sun, seems the place for a note on the use of mud, or at any rate earth, for building in Britain. Earth straight from the ground has been used—in various ways—more generally than most people realise as a substitute for brick and stone. Walls must be extra thick, though not necessarily as thick as the cob walls of Devon for which chalky earth was laid on in wet layers. Norfolk contains some clay-lump cottages, and at East Harling there was until recently a clay-lump council school in the Regency style—all with walls no more than a foot thick.

Nearly all counties can boast at least a few cottages or farm

16 *above right*
First century AD.
Columns of fluted
brickwork at Pompeii

below
One of several examples of
fifth-century brickwork at
Ravenna. In Italy
brickmaking persisted
after the disintegration
of the Roman Empire
in Europe

17 *far left*
Tenth century. Saxon parish church at Brixworth, Northamptonshire

above and below left
Twelfth century. St Botolph's Priory church at Colchester, Essex—note blind arcading; detail

buildings with mud walls, these being concealed beneath a most necessary covering of plaster and paint, tar, or weather tiles. Sometimes the occupants of old buildings are ignorant of what they are made of until they try to cut a new opening for a door or window.

A minor revival of the arts of building with earth took place during the brick shortages that followed the First and the Second World War. The architect Sir Clough Williams-Ellis gave encouragement during both periods and remains fascinated by the idea of getting houses to rise up literally out of the ground. He has probably written the last word on the subject, so far as its history goes, in his book *Building in Cob, Pisé and Stabilised Earth*, 1947.

Building with cob was the slowest and most laborious method, since each layer of wet material, laid on with spades, had to be allowed time to dry. Pisé de terre is the name for earth rammed down nearly dry between shutters: some experimental pisé cottages built at Amesbury in the 1940s remain in good order —another was added to the group in 1959. Stabilised earth consists of top soil and cement formed into blocks. Clay lumps resemble ordinary bricks in their green state—they are laid with a soft clay mortar. Sir Clough commented as follows on the matter of strength:

> All forms of earth construction have proved themselves adequate to support the loads of the normal two-storey house—about 1½ tons per foot run—provided walls of a generous thickness, usually about 18 ins. to 2 ft. are used. . . . It is claimed that compressive strengths of 1000 lb. per square inch and more can be achieved with cement stabilised earths, and while these figures may be somewhat extravagant, depending on careful control in the soil mix and very thorough compaction, it should nevertheless be possible to exceed 400 lb. per square inch, at which strength it would be reasonable to use the material in cavity

> construction with leaves 4 ins. in thickness, provided the usual number of wall ties were included. This would of course apply only to two-storey houses with joisted floors and with evenly distributed loading.

That, however, was written in 1947. Even experimental building with earth was virtually brought to an end in Britain by the coming of new materials and techniques, of easily available supplies of inexpensive bricks, and by the rising cost of labour which made it no longer an economical method of construction. The situation in less developed countries is different, as will be shown in the final chapter of this book.

4 Roman brick

Mud bricks were being used in Greece in the fourth century BC (they helped to make the walls of Athens), but in general they were for humble dwellings only and for the cores of stone walls. Mud bricks were later used in Rome, however, for the construction of substantial houses—bricks of higher quality than the Greek ones, according to the historian Pliny.

Burnt as well as sun-dried mud bricks were being made in and around Rome by the end of the first century BC, but Vitruvius in his treatise on architecture, written about 25 BC, refers almost solely to the sun-dried kind. His fellow Romans were at that time making burnt bricks as a matter of course in England, where the climate seemed to demand them; in the south these were reserved for such features on a mud building as a projecting course to throw off rain. Vitruvius, whose sections on brick make pleasantly soporific reading, discusses the details to be considered when making mud bricks:

> Gravelly, pebbly and sandy clay are unfit for the purpose; for if made of either of these two sorts of earth, the bricks are not only too ponderous, but walls built of them, when exposed to the rain, moulder away, and the straw with which they are mixed will not sufficiently bind the earth together because of its rough quality. They should be made of earth of a red or white chalky, or of a strong sandy, nature. These sorts of earth are ductile and cohesive, and not being heavy, the bricks made of them are more easily handled.
>
> The proper seasons for brickmaking are the spring and autumn, because then the bricks dry more equably. . . . When plastering is laid and sets hard on bricks which are

> not perfectly dry, the bricks, which will naturally shrink, and consequently occupy a less space than the plastering, will thus leave the latter to stand by itself . . . it soon breaks to pieces, and in its failure sometimes involves that of the wall. It is not, therefore, without reason that the inhabitants of Utica allow no bricks to be used in their buildings which are not at least five years old and also approved by a magistrate.

Mud bricks came in three sizes apparently: the didoron, a foot long and half a foot wide; the pentadoron, five palms of the hand each way; the tetradoron, four palms each way. The pentadoron was for public buildings and the other two for private buildings. Each sort had a half-brick made to suit it.

The Romans' first use for fired building material was to make tiles to waterproof roofs. But the builders found the tiles, and broken pieces of them, so valuable for strengthening vulnerable parts of walls that when, in the first century AD, burnt building bricks were demanded, these were made tile-like to match existing work—and also because thin bricks took less fuel to fire. The word 'tegulae' meaning tiles served also for bricks. The brickearth was sometimes trodden or banged out like pastry and then made into bricks by cutting.

A usual size was a foot or so square and $1\frac{1}{2}$ inches thick, but there were many others: small bricks to be laid on edge in floors (as in the Forum at Colchester), segments of all kinds for columns and octagonals for certain small columns. Walls were generally composed only partly of bricks, having them as a protective facing to a core of concrete made with lime and earth or with lime and volcanic ash.* In Rome itself there was a liking for pyramid-shaped bricks. The apex of each was embedded in the concrete, though occasionally the units would be reversed to produce a strongly textured effect. Since the concrete hardened slowly, the outward thrust of it during

* The Romans' discovery that volcanic ash near Pozzuoli made a natural hydraulic cement has led to the modern term pozzolanic.

construction was reduced by bonding courses about every three feet which spanned from one brick facing to the other.

Wherever it appears, Roman brickwork of the first and second centuries is technically and aesthetically at its best, the laying neat and the mortar joints no more than half the thickness of the bricks: these joints gradually widened until by AD 300 they might be as thick as the bricks.* An early second-century wall unearthed at No. 19 Tower Hill, London, gave a good idea of what Roman brickwork looked like when newly done. It was of Kent ragstone with, every few feet, two or three rows of flat red bricks running through and binding the rough stonework.

The kiln-fired bricks were made extensively by the Romans for work in the moist climate of England (and in the Germanic provinces) and were used for such military establishments as those at York, Chester and Leicester. Many were impressed—as Dr Norman Davey describes in *A History of Building Materials*—with a maker's mark in the form of letters, or a picture of an animal or a bird. Out of a large number of Roman bricks found in 1957 at Cottesford Place, Lincoln, some are stamped with the letters LVLE: the first three are believed to indicate the manufacturer and the fourth the particular kiln or works.

These markings, like those on Babylonian bricks, resemble modern practice whereby the bricks, for example, of London Brick Company are stamped with the letters LBC together with a number or letter identifying the works. Roman bricks have been found in London bearing the stamp P.P.BR.LON. The first P could stand for either Procuratores (financial officials) or Portitores (customs officers), while the rest of the letters seem to stand for Provinciae Britanniae Londinii.

The nature of Roman bricks indicates the employment of a very plastic clay carefully prepared and having a suitable admixture of sand. After they had been moulded or otherwise

* Some people admire wide mortar joints with thin bricks. In 1973 the architect Sir Clough Williams-Ellis was calling for rebated bricks of a pattern seen in Sweden which gave the effect of Roman tile-shaped bricks.

shaped, the soft wet bricks were left on a bed of sand to stiffen enough to be stacked in the kiln. During this period they might acquire unofficial marks: foot or paw impressions and light-hearted drawings of faces. A Roman brick in the Guildhall Museum, London is inscribed, apparently with a twig, AVSTALIS DIBVS XIII VAGATVR SIB COTIDIM, which means 'Austalis has been going off by himself every day for a fortnight'—thirteen days equals a fortnight in colloquial Latin. A worker seemingly wished to draw attention to the absences of one of his mates.

Roman brick kilns had several flues beneath the oven floor and were similar to kilns that had been in use two thousand years previously—and to some kilns of the Middle East today. In 1932 Dr Norman Davey excavated a Roman kiln at St Albans and gave this account of it:

> The structure, composed of pieces of brick and tile bonded with clay, was built below the natural level of the

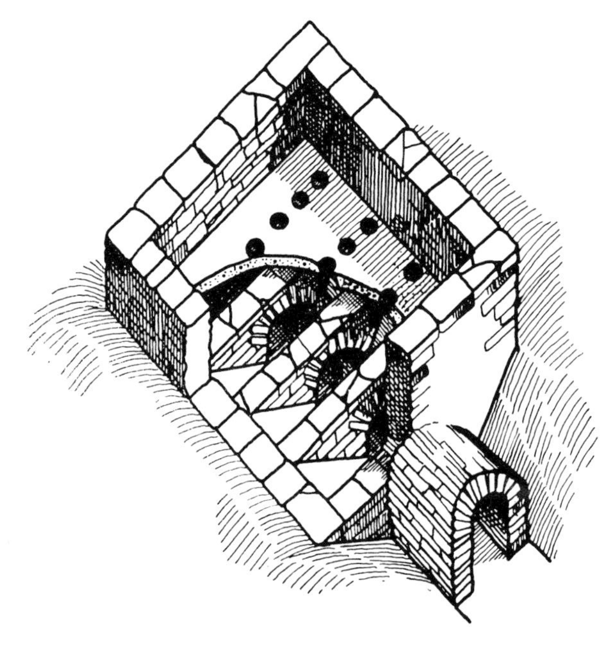

18
Roman brick kiln excavated near St Albans, Hertfordshire

ground. In this way the structure was solid and better able to withstand the stresses set up in it by the great heat, and the heat losses were greatly reduced. As the level of the oven floor was approximately the same as that of the ground, the stacking of bricks in the oven was easy. The kiln, as was usual, was built on the windward slope of the hill and the fire tunnel was lengthened to increase the draught . . . the products to be fired would have been surrounded and covered by pieces of burnt brick and tile smeared with clay to protect them from the weather and to prevent the heat escaping too quickly.

Although nearly all the bricks made in Britain by the Romans were red, in Rome the architects came to take an interest in getting them to emerge from the kilns in other colours. Among the extravagances of the final phase of the Roman Empire, in the fourth and fifth centuries, was a fashion for embellishing buildings with patterns of coloured bricks. They were worked

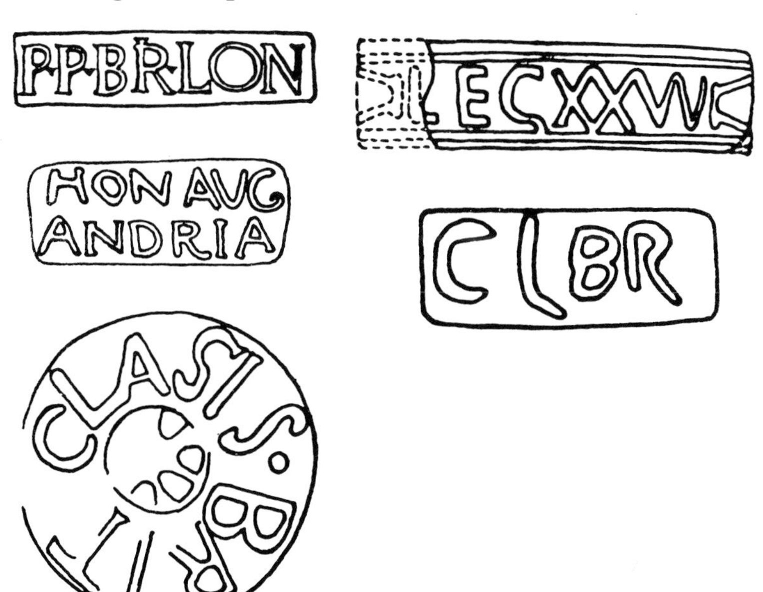

19
Official stamps on Roman bricks found in Britain

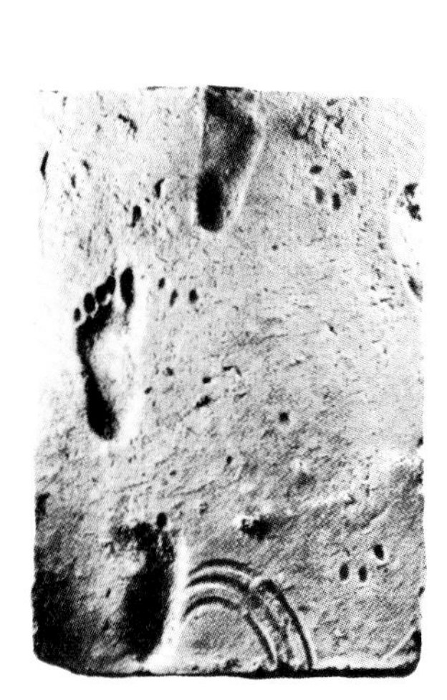

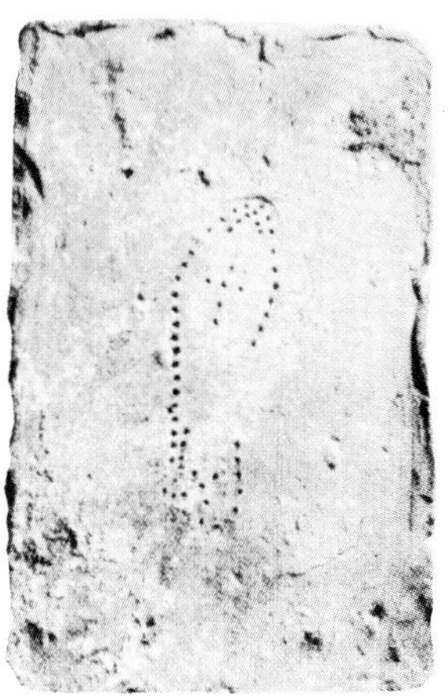

20
Unofficial stamps on Roman bricks found in Britain. People and animals trespassed on bricks left to stiffen on a bed of sand and occasionally someone drew a face on them

into the walls, among the classical features, to produce the sort of polychrome effects of yellow, red and brown which stamp many houses in Britain as late Victorian—these also, of course, were put up during the final years of a great empire. Examples of such Roman brick buildings still exist in Rome.

The craft of brickmaking appears to have ceased in Britain a little before the Roman legions departed in AD 412, and hardly a brick was made during a period of seven hundred years. But from the seventh century, when Saxon invaders had become established, second-hand Roman bricks were often employed for ecclesiastical and public buildings. The supply of excellent bricks to be prised out of ruins must have seemed endless: the Normans no less than the Saxons were able to make use of them, and today all over southern Britain there are modest churches of rubblestone in which Roman bricks appear the most substantial ingredient. Essex is the county with most examples.

At Holy Trinity church, Colchester, the Saxon tower is built of bricks taken from the Roman city of Camulodunum, the ruins of which lay to hand. Verulamium, another city, was systematically broken up late in the eleventh century (and again early in the twelfth) to get materials for what is now St Albans cathedral: the west front there has a profusion of tile-like bricks in conjunction with stone and is regarded as one of the best examples in Britain of re-used Roman brick. Of the parish churches there is an example of particular interest at Brixworth in Northamptonshire and at Bradwell (St Peter-on-the-Wall) in Essex.

Summing up, it can be said that though there are only a few places in Britain where Roman bricks can be seen doing their job in the buildings for which they were made, they still help to support plenty of buildings raised in later and less secure ages. The use of bricks for strengthening purposes, as demonstrated by the Romans, became part of the British building tradition. It is especially noticeable in chalk districts; the walls of flint which are common on the downs of southern England and in East Anglia are invariably laced with string-courses and quoins of brick.

Mediaeval and Tudor brickwork 5

New bricks and not old Roman ones helped to strengthen the flint abbot's lodging and guest house at Little Coggeshall in Essex, about 1190. They are rough in texture and dark red, and most are of the type known as the mediaeval great brick. This large flat brick, which commonly measures 13 inches by 6 by just under 2, was gradually ousted in the thirteenth century by units of handier size, though Waltham Abbey Gatehouse, also in Essex and built around 1370, has some great bricks that are 15 inches long. Great bricks follow ancient Roman example and in northern Italy they were being regularly used in the nineteenth century: they are especially noticeable in the parapet built on top of the massive town wall of Lucca.

Little Coggeshall's bricks, which archaeologists agree were produced on the site, have been hailed as the first of English make. However, bricks in the nave arches of Polstead church in Suffolk, *c.* 1160, so much resemble them that L.S. Harley can write in *Polstead Church and Parish*, 1965, of the likelihood that they, too, are English and ante-date the Coggeshall bricks by about thirty years—the walls of the church do contain other bricks and tiles that are undoubtedly Roman.

The Low Countries supplied the example and impetus for at last reviving brickmaking in England. On the Continent the craft had not ceased with the disintegration of the Roman Empire, and nor were the Dark Ages dark architecturally: in Italy the Byzantine basilica of San Vitale at Ravenna was brick-built in AD 547 and in north Germany, under Charlemagne, brick was used for Romanesque cathedrals from the eighth century.

Norman Davey notes in his *History of Building Materials* that the great brick occurs in northern Italy as early as the late eleventh century in the church of Santa Fosca at Tortello. He gives 'the very close intercourse in trade between the Venetians and the Flemings' as a possible cause of the northward spread of the craft. But already by the early twelfth century a positive renaissance of brick was taking place in Flanders, Holland and northern Germany. Builders and their patrons became especially enthusiastic about brick because large areas lacked stone.

The eastern counties of England were short of stone, too, and inhabitants enriched by trade were glad to take up new ideas brought across the North Sea by those Flemish immigrants who were brickmakers and masons. Flemish bricks came to be imported as well as made locally with the help of Flemish craftsmen—202,500 were shipped from Ypres in 1278 for work on the Tower of London—but such imports, costly in transport, occurred less often than has been claimed.

The advance of brickwork in eastern England, with the spread of such familiar features as the Dutch gable, was encouraged by merchants and sea captains who had to visit the towns abroad where brickmaking flourished and impressive buildings struck the eye. English brick could almost be called a by-product of dealings with the Hanseatic League: this was a powerful combination, formed in 1241, of trading centres in the Low Countries and north Germany. The town of Hull in Yorkshire, which became the first all-brick town in England, did much profitable business with the League. Even Hull's encircling wall and towers were made of brick. A municipal brickworks run by chamberlains was in operation between 1303 and 1433. The bricks it turned out, still referred to on paper by the Latin word tegulae (tiles), were mainly red and ranged in size between 11 inches by 5½ by 2 for the town walls, *c.* 1321, and 9 inches by 4½ by 2 for Holy Trinity church, 1315–20—this building was so artfully stuccoed in the eighteenth century that in 1756 a visiting antiquary supposed it not to contain a single brick.

21
Bricks of the small, Flemish type in the first post-Roman brick house in England: Little Wenham Hall, Suffolk *c.* 1280

The first dwelling house built with more brick than other materials is Little Wenham Hall, Suffolk, 1260–80. All the bricks are of the contemporary Flemish pattern, but since exactly the right clay for their colour—cream to greenish yellow—is to hand locally, it is no longer thought certain that they were imported. A large number of these Little Wenham bricks measure 9 inches by 4½ by 2: this is only fractionally different from today's British Standard Metric size of 215 mm by 102·5 by 65 (8⅝ inches by 4⅛ by 2⅝). In these times of rapid change, it is reassuring to notice that at least one artifact in common use has barely changed in shape since the thirteenth century, that the brick reached England seven hundred years ago as a perfectly developed unit. Apart from the great brick, rare after 1300, the brick as everyone knows it has varied in size by little more than an inch in any dimension.

Mr Harley, who is president of the British Brick Society, puts forward the theory that the measurement which has changed least is the half-girth. The breadth plus the thickness, he points out, are the important dimensions when it comes to picking up a brick; and he takes into account the smaller size of the mediaeval hand shown by suits of armour.

In the fifteenth century the main use of brick that developed was in the building of schools and colleges (Eton; Jesus College, Cambridge), churches and a few castle-like houses. Some of the more imposing of these brick buildings have a French air, the result of the men responsible for them having spent time among castles in France during the Hundred Years' War.

Herstmonceux, a fairy tale castle in peach-coloured brick, *c.* 1440, is certainly French in feeling: Sir Roger Fiennes, its creator, was at Agincourt. Now in the hands of the Royal Observatory with its beauty unmarred, Herstmonceux remained until nearly the end of the fifteenth century the only brick building of consequence to be seen so far west as Sussex—brick belonged to the eastern counties. Its tentative diaper patterns, formed with bricks having ends of a different colour,

22
As a result of the Hundred Years' War castles show a French influence. Tattershall Castle, Lincolnshire, finished in 1449

are an early example of a type of flat ornament in brickwork, used in France, which was to become common over much of England; the skill with which the patterns were worked in Tudor times—as at Farnham Castle, Surrey—may be the more marked if these are compared with modern versions. Caister Castle in Norfolk, whose bricks seem to have come from a site by the River Bure still called Brick Pits, was built around 1430 by Sir John Fastolf who had seen long service in France and, although Caister is now in ruins, its circular tower and corbelling show French influence.

But there were of course other Continental influences during a period when places like Lubeck, Ratzeburg and Danzig were famous for their Gothic architecture in brick. Tattershall Castle in Lincolnshire, completed in 1449 for Ralph Lord Cromwell (he also had served for years in France) shows a variety of influences, according to Dr D. Simpson in a British Archaeological Society paper (volume 40): these include the brick castles of Old Prussia and the Palace at Marienburg at Gdansk, Poland. Tattershall had moreover a Dutch master brickmaker, Bawdwin Docheman. The building must have seemed very un-English in its earlier years: today it looks like the prototype of romantic barrack towers in Victorian garrison towns and some universities.

In the Tudor period (1485–1603), brick became so well thought of that the rich were glad to have their houses built with it exclusively, even sometimes in parts of the country offering adequate supplies of stone. When people saw that brick was good enough for the king himself, its social acceptance as an alternative to stone was complete. A dark but mellow red was preferred; today the very word Tudor suggests the brick of that colour at St James's Palace and the older parts of Hampton Court Palace, and its use as a trade name for one of London Brick's facing bricks may have helped these to be best-sellers.

If Henry VIII's reign (1509–47) was the first great age of brick, it was also an age of remarkable—and pleasantly absurd—chimney building. London had had plenty of chimneys a hundred years earlier, but in the country as a whole (with outstanding exceptions like Tattershall and Layer Marney) anything more than a shaft of unfired clay was unusual throughout the fifteenth century. People put up with the smoke that failed to escape through a window or a hole in the roof. The purchase at Rickmansworth in 1425 of '2000 breke* for making chemeneys at Langley' was an event to be proud of—the quotation appears in the Rev. Valentine Fletcher's book *Chimney Pots and Stacks*, 1968. William Harrison, an Essex parson, reported for *Holinshed's Chronicle* (1577–87) that elderly villagers were marvelling and shaking their heads to 'see the multitude of Brick chimneys lately erected'. He adds: 'Now we have many chimneys, and yet our tenderlings complain of rheumes, catarrhs and poses'; in his opinion smoke had been a good medicine.

By the beginning of the sixteenth century most house-owners at least aspired to chimneys, and brick was widely called for as the best material for making them. It was also, since it readily took special shapes before firing, a material which allowed the brickmaker and bricklayer to outdo the stone mason in meeting a demand for show. Many *stone* houses were fitted from the start with brick chimneys. The bricks had to be bright red to make sure that the work stood out. Mr Fletcher reports that at Collyweston, Lincolnshire, in 1504 a deficiency in redness was helped out with

> vij lb. of red ocker with 1 oz. of the offales of the glovers lether, vijd. Item to John Bradley wiff for xiiij gallons of small ale for the said cheney of bryk, vjd.

The affluent appearance of the chimney stacks of Tudor resi-

*Bricks had formerly been called walltyles. The Norman-French word briche began to be used in the late fourteenth century and was spelt brik, brike, bricke or bryk. A fourteenth-century dungeon at Windsor Castle was recorded as being of petris and brikis (stones and bricks). Miss Jane Wight writes entertainingly of brick's vocabulary in *Brick Building in England*, 1972.

23
Early English brickwork at its best and mellowest at Hampton Court Palace, Middlesex, *c.* 1520. The chimney stacks with their purpose-moulded bricks may seem incongruous, but the ingenious designs and careful execution of this branch of Tudor work raised the prestige of brick.

dences, visible from afar, influenced those concerned with the building of lesser dwellings. Yoeman farmers added ornamental brick chimneys to their oak-framed houses and those who were building afresh knew that bricks were an urgent requirement, for the stack would go up first and have the rest of the house built round it. A surprising number of farmers, perhaps chimney snobs, contrived to have the upper part of their stacks made in the octagonal, hexagonal or spiral form of high fashion; work of this kind can often be seen where old half-timber buildings are grouped together. Famous examples of virtuosity in chimney building include those at Gainsborough Old Hall in Lincolnshire, Kirby Hall in Northamptonshire and Cobham Hall in Kent.

24
Italians were brought to England to form the terracotta transom windows and parapets of Layer Marney Hall, Essex, 1520

The use of brick for all purposes spread in the second half of the sixteenth century to several districts in which before it had been unknown. By 1574, according to the *Victoria County History*, Southampton had its official brickmaker, but he could not supply all that was needed for housing the gentry. Brick began to appear in the South Midlands and was even seen, mixed with stone, in Derbyshire and Lancashire. Contrary to popular belief, brick nogging was not yet being inserted in half-timber houses: there were plenty of daubers capable of making the wattle-and-daub infilling that put no strain on a wooden framework.

Long before the Great Fire, the advantages of brick were being appreciated in London by members of the professional classes. The founding of the Tylers' and Bricklayers' Company in 1557—early in Queen Elizabeth's reign—indicates the new importance acquired by brick. The Charter of Incorporation was confirmed by another granted in 1571* and further strengthened by charters signed in the reigns of James I and James II. These charters gave the Company full jurisdiction

* Talk of the so-called Elizabethan Statute brick has arisen through Nathaniel Lloyd supposing that this document, giving the brick size 9 inches by $4\frac{1}{2}$ by $2\frac{1}{4}$, referred to a statute where in fact it merely confirmed London by-laws.

over the manufacture of bricks and tiles for the City of London and also over the journeymen, bricklayers and tilers employed. The Company's authority, extending within a radius of fifteen miles from the City, included power to inspect finished and partly finished buildings, to impose fines for bad work and to bind apprentices.

Most people having some acquaintance with England can bring to mind Tudor buildings which are partly if not wholly of brick. They may not be well known; there are still large numbers of Tudor houses, in the Eastern counties especially. Because of rough handling and a very thorough firing, the slender and slightly curved and twisted bricks of Tudor times have an agreeable homespun look giving character to the walls they form. Where Tudor brickwork is oppressive in its redness, the fault may be found less in the bricks than in mortar joints that have turned black from town smoke—as at St James's Palace in London. But out in the country the wide pale joints never cease to tone down pleasantly the colour of the bricks. And if anyone still wonders why the joints were so fat, let him build a garden wall with a heap of old brick bats, some perhaps with cement attached, and try doing it with thin joints.

6 Early brickmaking practice

Pre-industrial brickmaking in England was a very local activity. In 1477 bricks to repair the walls of the City of London were made from earth dug in Moorfield, and the lawyers of Lincoln's Inn were able to make bricks from their Coneygarth next to the site. Wherever possible, people did their digging for brickearth in such a way that the cavities could be turned into moats or cellars.

Throughout the early centuries of brickmaking, the work—it was work for the summer months only—was done by builders and their labourers, by farmers and theirs and by itinerant brickmakers who were called in to produce the materials for a particular building. An agreement of 1644 relating to Syon House, Middlesex—M.W. Barley describes it in his *English Farmhouse*—gives an idea of what happened. John Hawkes of Hounslow was to have £5 towards digging the earth, £3 for each week when it came to working, £5 15s. at the end of making half a million bricks and, finally, 6s. 8d., for each 10,000 'well and sufficiently burnt and delivered out of the kiln'. As late as 1833 J.C. Loudon, in his *Encyclopaedia of Cottage, Farm and Villa Architecture*, advises farmers on how to make their own bricks; and many an isolated brick barn, by no means ancient, has near it a patch of irregular ground (there may be a small pond) which indicates where once there was a miniature brickfield.

Commercial yards equipped with permanent kilns existed none the less from the fourteenth century. One of the first was at Beverley in Yorkshire, where the town rented out land for brickmaking. Large estates such as the Duke of Bedford's (Woburn has a seven-mile brick wall) maintained their own brickyards. These were partly commercial in that they found a

25
Sixteenth-century brick-making in Germany, the text reading in part: 'All things I lay to my kiln, as a brickmaker, and prudently and with ease and skill I cook my bricks' —Hartmannus Schopperus, *Panoplia*, Frankfurt, 1568

market for bricks not needed on the estate: the Ashburnham Estate brickyard, founded in mediaeval times, became so well thought of that its products, fired with wood, were still being sold to the public in 1968.

Much of the charm of the earliest English bricks derives from their numerous imperfections. Carefree puddling and squeezing of the raw material, followed by fierce firing, produced plenty of misshapen specimens, a well-pitted texture and a variety of soft colours. The use of esturine clay may show itself in water-formed, wavy lines.

The bricks of the Cow Tower, 1380, guarding a stretch of the River Wensum at Norwich, were made of an incompletely stirred mix which fired to the shades of red, pink, yellow, grey and green. These bricks also vary in size, lengths ranging from 8 inches to 12, and widths from 4 inches to 7. It seems certain that nearly all of them were made by the method, often used by the Romans, of cutting them from layers of stamped-down clay. To prevent it sticking to the ground, the wet clay would be spread out on straw. The more general procedure, though, was to form bricks in moulds and then to arrange them on straw to dry. When a mediaeval brick is prized from its fellows in a wall an impression of straw, sharply etched, is often seen; and there may be lines showing where another brick had rested during drying. Examining these marks, and studying the documents of such a brickyard as Hull's, has shown students of brick that the method of making hand-moulded bricks has probably changed little since mediaeval times. Much study in this field has been done by the geologists Ronald and Patricia Firman, who reported results in 1967 in the *Mercian Geologist*, journal of the East Midlands Geological Society.

Brickearth was dug in the autumn, they confirm, and piled in heaps for the action of the weather, especially frosts, to break it down to some extent. In the spring it was tempered—that is, trodden or turned with spades to produce an even plastic consistency. The moulder, working at a table, sanded or wetted his wooden mould and threw into it with some force a lump of clay. He sliced the surplus from the top of the mould with a stick known as a strike or with a cutting wire on a bow.

The moulded bricks were carried—sometimes in the mould and sometimes between two small boards called pallets—to the drying ground and laid out in a herringbone pattern. When the first row was partly dry, a second was laid across it and when that was ready, a third, until a height of ten rows had been reached. This pile of drying bricks was known as a hack. For the bricks to be dry enough to fire might take a month, during which time straw covers protected the hack from rain and excessive sun.

Burning seems to have been mainly done in kilns of the simple up-draught kind. Otherwise, bricks were burnt in clamps (or heaps), but as the word clamp was often used of permanent kilns, there is uncertainty about the prevalence of the method. It is known, though, that kilns were made from unfired bricks if fired ones were not available and that green bricks were often

26
Bricks drying in hacks and so set that the air can pass between them—present-day practice in some British yards

stacked inside them in the way they were stacked in the hack. There are a few brickyards in England today where most of these procedures are still carried out.

Hull fired its brick kilns with peat turves, and on one occasion, according to Miss Wight, needed 84,000 turves to fire 35,000 bricks. But the normal fuel was wood and the Tattershall accounts of the 1450s record the cutting of hundreds of small trees for burning. Coal came into use for firing bricks in the seventeenth century. The firing in a wood-fuelled kiln, containing not more than 20,000 bricks, would be continued for a period of about a week.

Bricks may be regarded, geologically, as heat formed sedimentary rocks. Having so considered samples of mediaeval brick, Ronald and Patricia Firman were convinced from the evidence of plasticity, of fossils and other inclusions that only superficial clays were used; that no mediaeval bricks were made from the older carboniferous shales, Keuper marl and Jurassic clays, which are the raw material for about 80 per cent of modern bricks. This accounts, they say, for the obvious difference in appearance of mediaeval and modern brick buildings—and the reason for a less marked difference in East Anglia is that superficial deposits continue to be used there.

Nearly all mediaeval bricks were made of brickearth taken straight from the ground with nothing added. Exceptions include the bricks at Little Coggeshall. These have been found to contain a high proportion of coarse sand, all its grains about a millimetre in diameter, which is evenly distributed throughout a matrix of very fine particles. The Firmans concluded that the

> lack of intermediate-sized grains suggests that this is an artificial mixture. If this deduction is correct, then as early as the 12th century some brickmakers were aware of the value of adding sand to plastic clay to reduce the shrinkage on drying and burning. The technique was already known and used by potters in East Anglia.

Early bricks which are pale pink or yellow, as at Jesus College, Cambridge, were the products of earths which contained naturally a high chalk content. But around the middle of the fifteenth century brickmakers seem to have made an effort to avoid the lime clays and to concentrate on those whose iron oxide content ensured redness in the bricks.

Red was the colour aspired to by nearly everyone planning to erect a brick house or add a brick chimney to a timber house. When the redness was thought inadequate for a certain task, it might even be increased artificially. Records of a manor house at Collyweston, Northamptonshire, show that about 1505 the colour of bricks for chimneys was improved by a mixture of 'offalles' from glove leather and ale.

A picture of mediaeval brickmaking appears in a Netherlands Bible of 1425: it purports to show the Jews at work in Egypt but really shows contemporary practice. Another early picture—of a moulder at work—appears in a book of trades by Hartman-

27
An illustration purporting to represent the Jews making bricks in Egypt which actually shows fifteenth-century brick-making in the Netherlands —*Nederlandische Bijbel*, Utrecht, *c.* 1425

nus Schopperus published in Frankfurt in 1568. The earliest written picture of brickmaking is believed to be contained in a letter of 1683 from J. Houghton to the Sheriff of Bristol—it is quoted by Nathaniel Lloyd in his *History of English Brick-work*. Houghton describes digging up the earth in the autumn, weathering it in the winter, tempering in the spring, and the placing of dough-like heaps on the moulding table. At the unencumbered end of this table are

> boards nail'd about nine inches high to lay sand in and in the middle we fasten with nails a piece of board, which we call a stock; this stock is about half an inch thick and just big enough for the mould to slip down upon. Then we have a mould made of beech, because the earth will slip easiest from it . . . we also have upon the table before the mould a little trough, that will hold about three or four quarts of water which we put in, and in it a strike to run over the mould to make the bricks smooth. . . . When we are thus prepared with utensils, then one man strews

> sand on the table (as maids do meal when they mould bread) and moulds the earth upon it, then rubbing the stock and inside of the mould with sand, with the earth he forms a brick, strikes it, and pays it upon a pallat; then comes a little boy about twelve or sixteen years old and takes away three of these bricks and pallats and lays them upon a hackstead, a raised place. . . .

There follows a description of how the boy stacks the green bricks to dry, arranging them on edge in a herringbone pattern, and of how he covers the piles with straw; it took between three weeks and a month for the bricks to become dry and hard. Houghton remarks that a moulder helped by a temperer and by a boy to carry off the bricks, would make 2,000 'in a summer's day, viz., about fourteen to fifteen hours', though an extraordinary man could make 3,000. Without assistance, his day's output would be about a thousand. Alone or helped, his pay was 4s. for 1,000 bricks, a rate which remained roughly constant until the early twentieth century when it dropped by over a shilling. Houghton describes the firing:

> Our bricks being thus prepared, the next matter is to burn them, which is after this manner: When we begin a new brick ground, for want of burnt bricks we are fors't to build a kiln with raw bricks, which the heat of the fire by degrees burns, and this will last three or four year; but afterwards we make it with burnt bricks and we choose for it a dry ground, or make it so by making dreyns round it. This kiln we build two bricks and a half thick, sixteen bricks long from inside to inside and about fourteen or fifteen feet high; at the bottom we make two arches three foot high. . . .

After a rather complicated account of stacking the bricks in the kiln, Houghton writes of lighting a small fire in each arch to drive off the 'water-smoak' and of then increasing the size of the fires until the bricks become red hot throughout the kiln.

The firing would take no more than three days, including the cooling down.

> Then we sell the bricks as soon as we can for as much money as we can get, but usually about thirteen or fourteen shillings per thousand. The prices for making and burning is seven shillings the thousand, the wood three shillings the thousand.

By the second half of the seventeenth century the usual kiln was of the intermittent type known as Scotch. It was a large chamber, open at the top, with a series of fire holes along each side opposite one another. The dried bricks were stacked inside in such a way that the hot gases could surge up between them. A layer of old burnt bricks was then spread over the top, the ends of the kiln were bricked up and roughly plastered over with clay, and fires were lit in the fire holes. When the burning process was completed, the kiln was allowed to cool and the bricks taken out. Under this method of manufacture, the shade of the bricks varied a great deal, those nearest the fires being the darkest. These harder-burnt, darker bricks had also shrunk more. On being taken from the kiln, the bricks were roughly sorted to give the purchaser consistent lots.

But regular kilns were put up only in brickyards with some degree of permanence, and firing in open clamps was probably more usual. This kiln-less method was favoured by the itinerant brickmakers, whose equipment on coming to a site need be little more than spades and a wooden mould; they reckoned to find their water and fuel locally. Bricks were now considered so necessary in many parts of Britain that commercial brick-yards were multiplying and, on the outskirts of towns likely to expand, some quite large works were set up. Many parishes, too, had their own brickyards. Whether a builder applied to one of these places for bricks, or made his own—perhaps with the help of an itinerant gang—depended on the cost of transport by horse and cart.

28
Late nineteenth-century pugmill made in Sheffield

A labour-saving piece of equipment which emerged at the end of the seventeenth century was the pugmill for blending the clays (or pug). The pugmill was a kind of barrel containing a vertical shaft with projecting blades. There was a long beam attached to this shaft to which a horse could be harnessed to give the motive power by walking round and round. The raw materials were fed in at the top of the mill and came out at the bottom as a reasonably smooth paste. A little girl's experience with a pugmill near Ramsey late in the nineteenth century appears in Sybil Marshall's *Fenland Chronicle*, 1967:

> When I was very small, my mother worked in the brickyards. I had a small swing fixed up for me on the beam opposite the horse, and round and round I used to go all day. If the horse dropped any dung, it had to be cleared away immediately, to keep the path from becoming greasy and in bad condition. So, as I travelled round, I kept a diligent watch for this, and was delighted when I could call out, 'Tom, old Jack's messing again.'

Brickwork of the seventeenth and eighteenth centuries 7

The Great Fire of London in 1666 was the more devastating for the fact that the majority of houses were still wooden. The City authorities insisted afterwards that new work should not be done with anything so combustible as wood and strongly recommended brick, described in the Act for the Rebuilding of the City of London as 'comely and durable'; not only outside walls were to be built of brick, or stone, but also all ornamental projections.

29
Seventeenth-century brickmaking in the Netherlands. Note the Dutch gables, which were copied in eastern parts of England—J. Luykens, *Trades and Professions*, Haarlem, 1695

A huge demand for bricklayers had the effect of undermining the power of the Tylers' and Bricklayers' Company: the number of craftsmen permitted to work as freemen for the Company (and for other City Guilds) was inadequate for the emergency, and workmen were liberated by Act of Parliament from the obligation to join a guild before practising their trade. The Tylers' and Bricklayers' Company never regained its authority, though a recoup of finances in the nineteenth century has allowed it to hold its own among the City Companies in social and charitable work.

Fire caused ruin in several other towns, among them Marlborough in 1653, Northampton in 1675 and St Ives in 1680, and each was rebuilt largely or wholly in brick even where stone quarries were near. Brick was not only cheaper but, at this period, no less fashionable. Mr Barley says in *The English Farmhouse* that there was enough demand for bricks in the Isle of Axholme, Lincolnshire, to encourage men to make them anywhere, anyhow. William Occarbie of Crowle was presented to the manor court in 1649 'for making bricks on the common and selling them out of the manor'.

The London building regulations did not affect the provinces,

but in so far as they offered good ideas for protection against fire, they were widely taken into account. A London Building Act of 1707, forbidding the use of the wooden eaves cornice, led to its replacement with a brick parapet. In 1708 it was forbidden for wooden door and window frames to be fitted unless they were set back 4 inches from the face of the wall; and wooden beams, no longer allowed to serve as window lintels, were largely replaced by the brick arch. It may be seen in old buildings today that these ideas were followed in due course many miles from London.

30
Thirteenth century. Little Wenham Hall, Suffolk, c. 1280; the first house in England with more brick than other material in its walls. The bricks were apparently made on the spot by Flemish immigrants who began to settle in East Anglia in the twelfth century

It was common in the seventeenth century for brick buildings of any consequence—farm buildings are certainly included—to sport such Dutch elements as curved or pedimented gables. These are readily seen near London as well as in East Anglia. Obvious examples, influential in their time, are Kew Palace, 1631, Broome Park, Kent, 1635 and Raynham Park, Norfolk, 1635.

Sir Christopher Wren, architect of St Paul's, liked to employ brick to achieve more than one effect. For some buildings he favoured rough Tudor bricks generously packed round with mortar and for others rubbed bricks of such regularity that joints were threadlike. Examples of the latter kind of brickwork are to be seen in his entrance to the Middle Temple and in the Fountain Court of Hampton Court, where smooth red brick, almost invisibly jointed, serves as a field for big pilasters. For this refined type of brickwork he specified extra small bricks, about 6 inches long by 2 inches deep. Where others were prepared to make brick a vehicle for architectural ornament, Wren and his followers used stone with brick as the background.

By Queen Anne's accession in 1702, brick was becoming a familiar building material in all parts of England to the east and south-east of the limestone belt. Here and there even labourers' cottages were built of a material valued for the way it kept out, as wattle and daub did not, the plague-carrying rat. Seventeenth-century building technique was much in-

31
Fourteenth century. The pack horse bridge near Bury St Edmunds. Made of brick only, it is believed to be the only bridge of its kind in East Anglia

fluenced by the bricklayer, settling wherever supplies of clay were turned to his use. Independent of the ancient art of the stonemason; he was familiar with new building methods and knew how to put his easily manipulated bricks to a variety of uses.

In Queen Anne's reign brick that was red, embellished where possible with dressings of Portland stone, held undisputed sway. The alternative to stone for dressings, especially those emphasising the quoins of a house, was a special crimson brick laid flush with thin joints for which putty instead of mortar might be used. This 'gauged' brickwork was employed, too, for window arches and jambs for string-courses and, later on, for the pilasters of more important houses; it went well with the duller red and the wider joints of the rest of the façade. From the middle of the eighteenth century, however, some people said the varying shades of red against white windows—and often white stone—was over-pretty in the sharp contrast produced. Stone, surely, was showing off the brick where the reverse would have been more seemly?

Comments on the subject by Isaac Ware in his *Complete Body of Architecture*, 1756, have never ceased to be quoted. He wrote of a red-brick greenhouse in Kensington Gardens (which he considered well-designed) that

> the colour is fiery and disagreeable to the eye; it is troublesome to look upon it; and in the summer it has an appearance of heat which is very disagreeable; for this reason it [red brick] is most inappropriate in the country.

If by 'in the country' Ware meant 'against a green background', what he had in mind is supported by a law of nature that since red is the complementary or directly contrasting colour to green, only a little of the one goes with the other without a clash. Thus a red telephone box can cheer a village green where a large, perfectly red building would be an affront—and who would choose a plain red sofa to stand on a plain green carpet?

32
Eighteenth-century brick-making in northern France —*Descriptions des Arts et Métiers*, Paris, 1761

Only some of the non-Oxbridge universities are red brick. In 1955 Sir Basil Spence stood on the rolling Sussex downs near Brighton considering the colour of bricks for his new Sussex University complex. 'Certainly not red against all that green,' he remarked to Redland Brick executives who were with him. It was decided their hue should be the russet of an old leather pocket book that he had on him. Redlands made the bricks specially.

With brick, of course, bright redness can be reduced by wide joints of pale mortar. But in late eighteenth-century London redness of any kind became unfashionable: instead of a contrast with greenery or stone, there was to be the harmony provided by grey stock bricks. Red ones were used only for dressings. By the end of the eighteenth century the change in taste had altered the face of London.

Although the bricks were generally described as grey in specifications, they nearly always turned out yellow-orange or yellow-brown and, before soot and smoke dulled them, eighteenth-century London houses were the colour of the recently-cleaned interior of Charing Cross Station. To get the preferred colour it was necessary to add finely ground chalk to brick-earth which would otherwise have fired red, but this was easy enough for the local brickmakers of London, since a layer of chalk occurred below the superficial clay deposits which they used. The chalk was useful, too, in that it reduced the contraction of the clay in firing and acted as a strength-producing flux.

What London builders were doing others tried to copy, and tones of ochre, sulphur and buff were extolled, especially by admirers of Palladian architecture, as the only proper colours for a brick house. Gault clays burn yellow, but more often than not the materials for making yellowish bricks were not to be had. Thomas Coke was able to avoid any obvious redness when building Holkham Hall in Norfolk; this great house is a pleasant brownish-yellow, having a little red brickwork only in the inward-facing walls of the courts which few people see. Some

33
House known as The College—Ashford, Kent. One wing received a Georgian brick front (and side) in the eighteenth century; the other has its sixteenth-century wall still uncovered

owners of red-brick country houses took the trouble to have them faced with yellow mathematical tiles simulating brickwork: these are described in chapter 19.

In the brick-building areas of England as a whole, however, there was not the least objection to red brick, though the colour was sometimes softened in the south-east by the inclusion of blue headers. Brick of one shade or another of red was the main building material for the symmetrically-fronted houses of the merchant and professional classes in cathedral cities like Chichester and Salisbury, market towns like Newbury and Devizes and ports like Bristol, Bridgwater and Portsmouth.

The familiar Georgian elevations, with a parapet half or wholly hiding the roof, fitted in with London by-laws of 1707 (the result of continued anxiety over fire) which made it illegal to have naked wood at the eaves. Numerous timber-framed houses in the provinces had their faces mutilated to take a casing of brickwork pierced for classical windows and doorway, while their side walls might be merely tile-hung and the back wall left alone. The job was done much less because of risk of fire than to be in the fashion. Indeed the urge in the eighteenth century to make old timber-framed houses look modern became an obsession; old-looking buildings were only good for cottagers, paintings and students of the Picturesque.

Eighteenth-century house-owners took even more trouble to hide timber than twentieth-century house-owners to expose it. In a provincial street today one house after another may appear Georgian, to be revealed as mediaeval or Tudor only by a fragment of roof, the design and placing of a chimney or, to the determined observer, by the back of the house and the architecture within. There have been instances of preservationists from the Georgian Group being briefly tricked.

The swing of taste away from the existing oak houses is reflected in the recurrent phrase 'new Hansome Brick House' in mid-eighteenth-century property advertisements. But for the poorer buyer, just a brick front would do. Mrs A. Walker supplies this example from the *Ipswich Journal* of 24 September 1748:

34
Dignity by Robert Adam in handling large masses of brickwork—Mersham-le-Hatch, Kent, which contains 2½ million bricks

> TO BE SOLD.
> A Hansome commodious House,
> containing six rooms on a floor with a
> Brick Front and Sash'd Windows. . . .

In 1784 Pitt the Younger introduced a brick tax to help meet the expenses of the American War of Independence; it was payable by makers following counts of such green bricks as were on view in their yards. The starting rate of 2s. 6d. per 1,000 went up to 4s. ten years later, and in 1803 to 5s. for the smaller bricks and 10s. for the larger. Since, to start with, bricks of all sizes attracted the same tax, there was the inevitable result that bricks were made tax-beatingly large.

There are specimens of the new, eighteenth-century great brick to be seen at Ware and villages nearby in Hertfordshire. One brickmaker turning out bricks twice the normal size was Sir Joseph Wilkes whose yard was at Measham in Leicestershire. His large products, referred to by Kenneth Hudson in *Building Materials*, 1972, were known locally as Wilkes's Gobs; a row of cottages built with them and called The Brickyards can still be seen. There are other examples of building with gobs, Mr Hudson says, in Measham and Ashby-de-la-Zouche. A member of the British Brick Society supplies further scattered examples. At Horncastle in Lincolnshire, he states, a house in West Street, built in the 1780s, has bricks that measure 11 inches by 5 by 3¼. The Weir House at Bodenham in Herefordshire went up forty years before the brick tax, but a screen wall on each side, which was added during its operation, contains bricks laid in English garden wall bond that are 12 inches by 6 by 3¼: these indeed represent a re-flowering of the mediaeval great brick.

The increase in size was shortlived, for an Act of 1803 put a double duty on bricks of more than 150 cubic inches—bricks measuring exactly 10 inches by 5 by 3 are likely to date from this time. The brick tax, which was repealed in 1850, encouraged to some extent the hanging of plain tiles on walls, for roofing tiles remained untaxed throughout the eighteenth century; and in rural parts it led to a heavy run on wood for cottages, especially for making weather boards. By the end of the century brick was nevertheless nearly always a cheaper building material than stone, despite the tax.

The bricklayers and other craftsmen of the great brickmaking regions were taking as much pride in their work as the mediaeval masons. It seems that there was a great desire to demonstrate that anything the stone mason could produce, they could, too; there are buildings in which even rusticated stone quoins are imitated in brick. But the Georgian way of assembling patterns with over-burnt headers, even in the walls of quite humble buildings, owes nothing to the tradition of the stone mason.

In America not only were the brick buildings of the seventeenth and eighteenth centuries largely influenced by British practice (Wren himself designed Wren House, Williamsburg), but nearly all the bricks, according to Charles T. Davis in his *Treatise on Bricks*, were until the late eighteenth century imported from England: they came over as a profitable ballast on vessels which had light cargoes. However, America's very first burnt-brick houses of consequence were gabled and thoroughly Dutch. They were built on Manhattan Island in the 1630s, with bricks from Amsterdam, for the occupation of the governor, Wouter Van Twiller, and of his staff.

Wood, which the English colonists knew how to use, has always been in North America a more usual building material than brick, or stone. But even in early days brick suggested wealth and taste—and it still does, to judge by all the wooden houses and shacks in the United States which have been given a brick veneer. It was considered a choice building material from the beginning in the state of Pennsylvania. Three years after founding the colony there in 1662, William Penn wrote as follows to his agent on behalf of a woman who proposed to emigrate but did not wish to live in a wooden house:

> She wants a house of brick, like Hannah Psalter's in Burlington, and she will give £40 sterling in money and as much more in goods. It must have four rooms below, about 18ft by 36, the rooms 9ft high, and two stories height.

One of the first public buildings made of brick in America was the old court house of Philadelphia, 1705. Looking like a Queen Anne town house in England, it stood in the middle of Market Street and was in constant demand for one purpose or another, Davis writes, until demolition in 1837—with bricks at 29s. 6d. per 1,000, the total cost had been £616. Another early brick building in Philadelphia was the Great Meeting House of Friends on the corner of Second and Market Streets, 1695.

Wood was the obvious building material for the Puritan emigrants at Boston, but when the wooden Towne House of 1657 was burned down in 1711, it was rebuilt with brick. Boston's once renowned Triangular Warehouse, at the head of the town dock, was built by London merchants in brick around 1700 on foundations of stone.

The manipulation of brick 8

Anyone building a brick wall instinctively gives it coherence by overlapping the bricks and avoiding the occurrence of vertical joints immediately over one another. Mediaeval bricklayers in Britain did this in a rough and ready way which is today dignified by the term haphazard bond, but by the fifteenth century it was widely realised that a regular bond both looked better and was marginally stronger.

Where the units of early brickwork are manipulated according to a system, it will be found that this is always English bond, which consists of courses of headers (bricks having their ends to the wall face) alternating with courses of stretchers (bricks having their sides to the wallface). This bond is robust and usually very easy on the eye; only in a few modern examples, where the tone of the mortar contrasts markedly with that of the bricks, is there a certain stern monotony in the courses of headers and stretchers and the way they emphasise the horizontal lines.

English bond was first conscientiously employed in England at Tattershall Castle, Lincolnshire, finished in 1449. It is seen in mediaeval French castles and was probably imported from France. This bond remained standard until the seventeenth century. Then Flemish bond was introduced to England and became the more admired of the two: it was used for the walls of Kew Palace and of Raynham Hall, Norfolk, both of which were erected in the 1630s. It became almost universal in Queen Anne's reign.

Flemish bond comprises alternate headers and stretchers in the same course and has never ceased to be a favourite for looks. It gives a feeling of unity because the brick faces clearly retain

35
Seventeenth-century manipulation of brick at The Dutch House, Kew, 1631. Note the rustication in which the windows are set, giving the contrast of apparently larger units, and the use of Flemish bond which at this date was an innovation

their individual shapes, one of which being half the size of the other, and because each header is related to the corresponding stretcher resting centrally upon it. Flemish bond always appears appropriate for buildings in which the walls take, or appear to take, the full weight of floors and roofs. Thus in the red damask-like walls of Wren's part of Hampton Court Palace, the separate bricks seem, in the phrase of P.M. Stratton, to have been stitched into a design, becoming part of the minute workmanship which sustains the whole: stone cornices, entablatures, pillars and frames stand out from brickwork which holds them in its bond.

So good an arrangement as Flemish bond was bound to inspire variations. A not unfamiliar example is Sussex bond, sometimes called Flemish garden wall. This consists of two, or three, stretchers to one header in every course: the headers are axially over one another alternately with stretchers. Sussex bond is easy on the eye in a long garden wall unbroken by voids: it was not used for Georgian façades because it limited freedom in window spacing.

English cross bond is a slight deviation from pure English bond in that the second brick of alternate stretching courses is a header. As a result, the stretchers break joint and the pattern is less mechanical. Dutch bond achieves similar diversity: the first stretcher course begins with a three-quarter brick followed by a header while the order in the second stretcher course is three-quarter brick followed at once by stretchers. English garden wall bond has a header course laid over three courses of stretchers.

Construction with headers only, still done for circular work, is thought by many especially pleasing: straight walls have a three-quarter brick to start each course. There are numerous examples at Lewes and in other parts of south-east England.

For the Old Hall at Ormsby St Margaret in Norfolk (*c.* 1735) the whole elevation was done in header bond. The multiplicity of joints in this bond gives a unified texture, the whole suggesting slow careful work carried out for the sake of an easy effect. This can be compared with the effect achieved with squared flints at Goodwood House in Sussex, where the smallness of the units brings the charm of mosaic or woolwork. Header and English bond come in the same class because of the feeling of strength in the compact pattern.

What most encouraged eighteenth-century builders in header bonding was a desire to make use of bricks having one end coated with an attractive blue-grey film. Huge numbers of these emerged from the wood-fired kilns as a consequence of wood smoke and the way bricks had to be set in such kilns.

Modern buildings commonly show nothing but stretcher bond except for a header at the end of each alternate course. This is economical in bricks and labour and the obvious way of laying

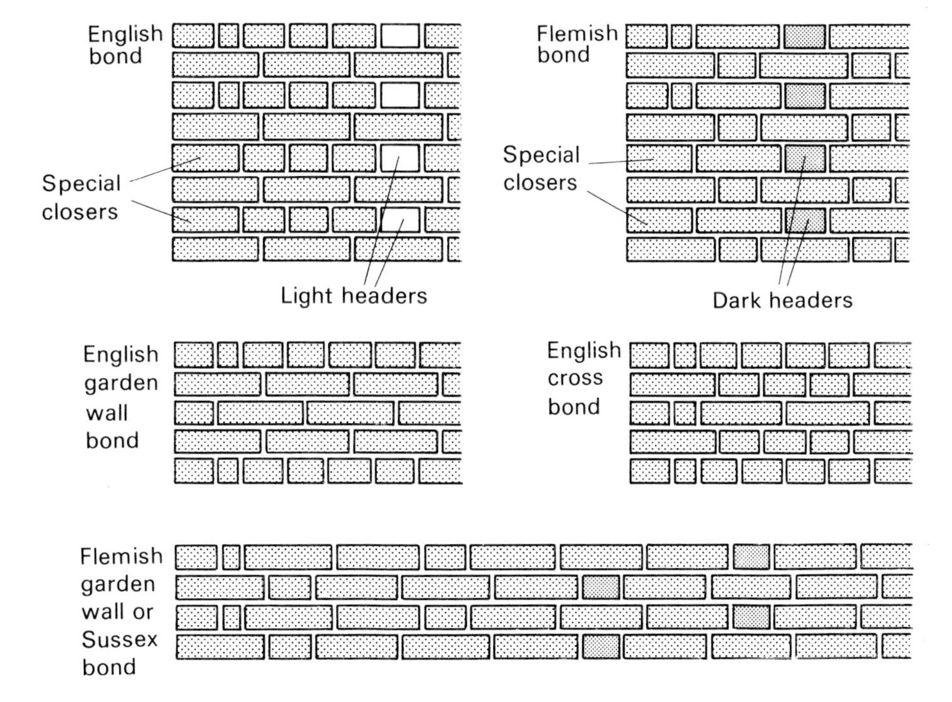

36
The more common brick bonds

the now standard 11-inch cavity walls of small houses—that is, walls with two 4¼-inch leaves having a 2¾-inch gap spanned by metal ties.

Although stretcher bond generally appears dull and monotonous, the only alternative in a full-cavity wall is to create just the look of, say, Flemish or English bond by the use in the outer leaf of half bricks. Late Victorian and Edwardian builders commonly took the trouble to do this for the better houses. There was also in those times the semi-cavity wall in which headers span an extra narrow cavity and serve as ties. Time has proved that the semi-cavity wall, well laid, can be effective for keeping out water. But this construction no more comes within approved modern practice than the solid 9-inch wall, and seemingly walls of bricks in stretcher bond will continue to proliferate, tending to degrade brickwork in general.

It must be said, though, that the poor effect of relentless stretchers is much reduced where the mortar joints strike the eye as little as possible. Certain red bricks are a problem, of course, since mortar that is reddish or dark in tone looks contrived, but at least excessive whiteness (and contrast) can be avoided. Where buff or grey bricks are being employed, these blend well with mortar of a normal mud-to-putty colour. An increasing demand for 'natural-looking' bricks is to be applauded, for it cannot be denied that some of the new small houses built with them—and having an overall hue of stone—can fit well into most landscapes, even in places where stone building is prevalent.

Any careful consideration of brick bonds raises sooner or later the question of why we should bother about the pattern produced when the clean white or colour-tinted surfaces of stucco are easily attainable. The first answer is that stucco needs continual maintenance to keep it clean and uncrazed—more in Britain than in the hotter countries where it has become traditional—and the second that in a land of good cheap bricks there is no sensible reason for covering them up.

Stucco or its equivalent has often proved welcome as a grand fur coat on large national and commercial buildings, but for these there is generally plenty of money to spend on maintenance. Only careful and regular attention allows the Nash terraces of Regent's Park, London, to look as they do, for the stucco of these Regency-classical buildings lacks the endurance of Portland stone for which, in order to economise, the architect John Nash made it the substitute. It is slightly ironical that the great London Brick Company has its headquarters in one of these covered-over brick houses.

As moulding and modelling become more and more merged into flat surfaces, so bond pattern seems especially desirable. Bricks and stones are small units of construction whose appearance and obvious function adds something to the whole effect: consider the drabness and stains of so many unbroken surfaces of concrete.

The way of doing mortar joints is as noticeable as their colour. Pointing the surface with an extra hard cement mixture is no longer in favour for new work (its purpose was to contain weak mortar) and the flush joint, formed by chopping off surplus mortar as the work goes on, is considered both workmanlike and satisfactory for the presentation of bond, neither exaggerating nor diminishing it. Raked out joints to emphasise the bricks is largely a north European practice.

Mortar looks best if its texture matches that of the bricks. To this end a coarse sand is useful, though bricklayers have various tricks for roughening their joints: smoothing them

37
Types of mortar joint

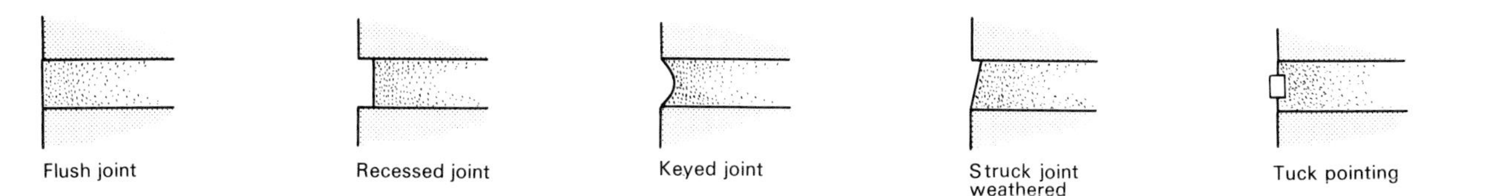

out with the point of a trowel seems appropriate only in certain engineering work. The weathered joint, cut in below the upper brick and sloped out to the face of the lower brick is still thought by some to throw water off a building, but the vertical joint does not mitre with the horizontal and the undercutting of the brick courses gives a hard effect against the light-catching slope of mortar below. The purpose of the now obsolete tuck joint, from which a narrow band protrudes, was to get the last ounce of effect from common bricks used for the early suburban houses.

38
Fifteenth century. The bricks vary considerably in all dimensions at Caister Castle, Norfolk, 1435. Such wide variations make it difficult to date early work by sizes alone

Brick, as an artificial substance, offers easier opportunities than stone for manipulation for the sake of ornament. Baking on glazed pictures to spread across an area of brickwork and utilising bricks of different hues went on thousands of years ago in Babylon and Persia. Wall patterns with coloured bricks made a noticeable appearance in France in the fifteenth century: near Rouen an elaborate octagonal pigeon house of red, yellow and green bricks) which may even be late fourteenth century) was recorded by J.L. Petit in the *Archaeological Journal*, volume 9, in 1852. During the past hundred years decorated brickwork in northern France has repeatedly gone too far in the direction of what V.S. Pritchett has called the coloured crossword puzzle. In Britain there has been less showiness—except in exuberant nineteenth-century creations.

Even when the eighteenth-century builder was making on the site only enough bricks for a particular undertaking, he would have a choice of colours to play with because not all bricks came out equally burnt. In Victorian times builders and designers of buildings took advantage of different coloured bricks to make patterns which are uncomfortably strident, but the Georgians appreciated that there was no need for strongly contrasting shades. Their patterns were usually made with headers, especially the blue headers already described. The greatest variety of such work is to be seen in south-east England, and those wishing to read about it in detail will find

39 *above left*
Fifteenth century. Diaper patterns on stone-dressed walls of Morton's Gateway, Lambeth Palace, London. 1495

below
Wall with plenty of headers at Penshurst Place, Kent

much of interest in Alec Clifton-Taylor's *The Pattern of English Building*.

Many other ways of laying bricks to give an ornamental effect have been followed. Sixth-century Italian builders achieved a pleasant saw-tooth band by laying ordinary bricks corner to corner (this was popular for wall tops throughout the eighteenth century in Britain), and they liked to break up flat surfaces with blind arches and small pendant arches. The herring-bone brickwork of fireplace surrounds was once favoured for wall panels, especially as nogging to fill the spaces between the members of a timber-framed building. The ornamentally laid brick nogging seen in timber houses has often replaced an original filling of wattle and daub. Such replacements have been made since the end of the seventeenth century, but not always with perfect results because, as well as tending to be too heavy for the structure, brickwork panels can introduce dampness by projecting a little from the wood and allowing rainwater to seep in at the joints.

Various raised patterns can be produced quite easily with ordinary bricks. The strapwork popular in the seventeenth century consisted of bricks laid with a projection of about an inch to form ovals and squares. At different times bricks have been arranged diagonally to give a serrated edge, or every other brick in a row of headers has been set sticking out to form a dentil pattern.

Another vogue—of the eighteenth century—was for rustication, which meant so laying the facing bricks of an elevation that they formed projecting blocks resembling stone. Similar projections for dressing purposes have been formed with stucco, but the ingenious manipulation of bricks, whereby their smallness is played down to give an idea of boldness, can be especially pleasant to look at. Rustication in brick is at its best with a roof of rugged stone slates.

So far only standard bricks have been considered for ways of enriching a surface. But for centuries bricks have been given special shapes to serve the demands of architectural detail. Brick clay can easily be purpose-moulded; the fired material can be carved, cut with an axe, or rubbed with abrasives or with another brick.

Bricks to be rubbed to a smooth finish are made softer than usual by increasing the proportion of sand to clay in the mix and by carefully baking rather than burning in the kiln. Known as rubbers, and perfectly durable, these bricks were constantly in demand in Georgian times for such dressings as pilasters, window surrounds and arches.

Towards the end of the eighteenth century, house builders were fastidious about their brick arches and had them made with slightly wedge-shaped bricks instead of just thickening the mortar joints of ordinary bricks towards the outer radius. Every brick had to be cut with a saw to the required shape and rubbed down to make the finest possible joint: such brickwork is known as gauged.

The sculptural carving of brickwork was formerly undertaken only on rubbers, of which a panel or panels would be set in a wall. Work of this sort was in the province of the trade carver or brickmason until the 1920s and 1930s when a new breed of sculptor became interested: Pepys Cockerell carved a hunting scene on a garden wall at Haslemere and Eric Kennington the reliefs on Stratford Theatre; Eric Gill carved a crocodile in the facing bricks of the Mond Laboratory at Cambridge and, with Anthony Foster, a St Andrew over the porch of the Roman Catholic church at Gorlestone.

These facts are taken from an article by Mr Walter Ritchie in the *Brick Bulletin* of November 1972. A pupil of Eric Gill, he works direct on several materials and has executed remarkable brick carvings on churches, schools and a fire station. The advantages of carving a wall instead of a panel is the freedom to adjust, he says, and also the feeling that comes of identity with

40
Walter Ritchie sculpting low relief in brickwork consisting of wire-cuts set in mortar that is one part cement to three parts sand. The result was put on view by the Brick Development Association at the 1973 Building and Construction Exhibition in London where it suggested a new slogan, 'Brick is Love'

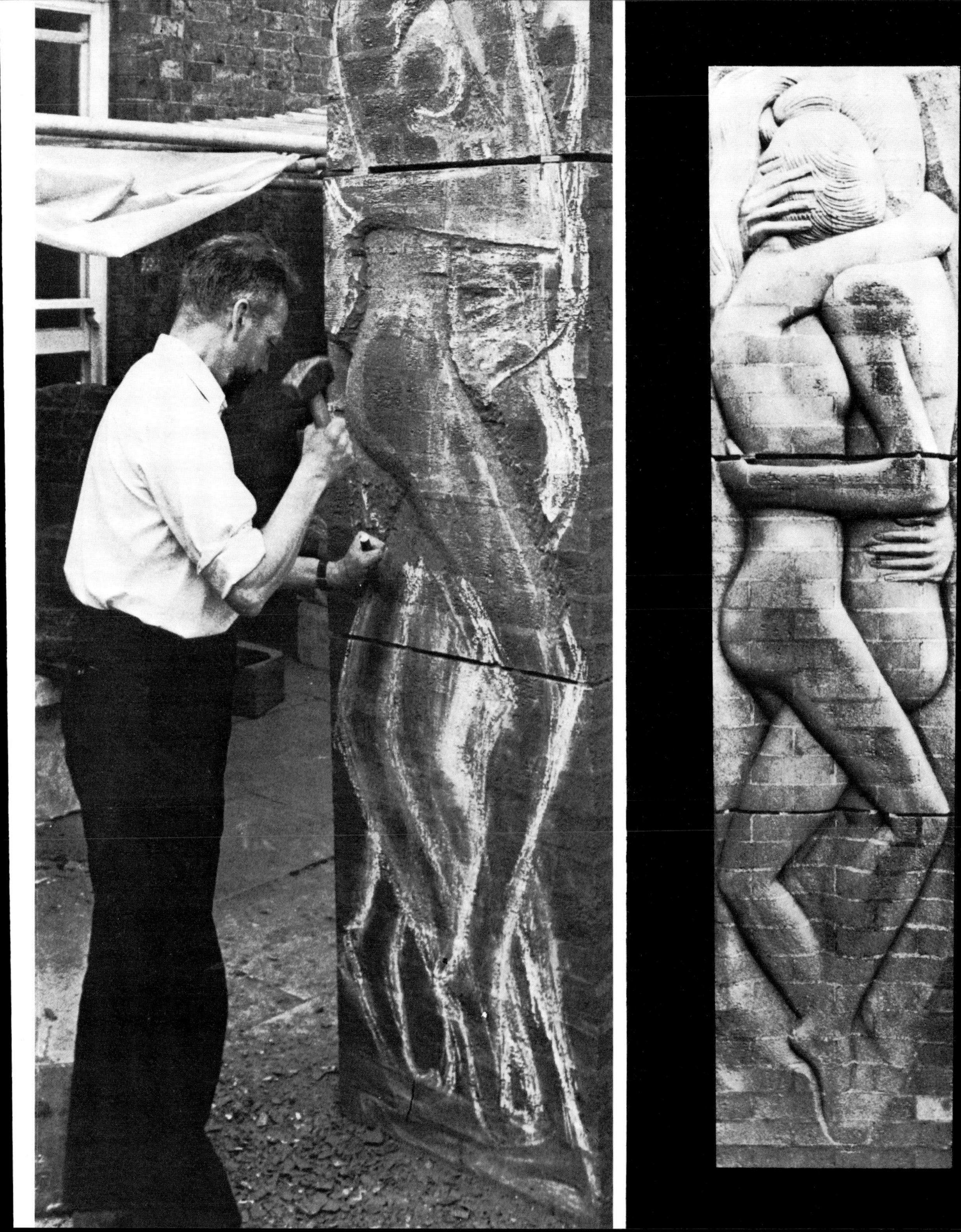

the building. 'The milder facing bricks can be a pleasure to cut and take detail as well as fine grained limestone and certainly hold it better in an acid atmosphere.'

Proper carving, going perhaps 3 inches deep, needs plain bricks without perforations, or the depressions called frogs, and backed up solid. 'It should be possible', says Mr Ritchie, 'to cut through brick and joint without change of direction or appreciable variation in consistency. . . . The grid of the mortar joints can be emphasised or reduced, but inevitably dictates a bold treatment.' His comment is certainly borne out by an abstract sculpture on an outside wall of the International Building Centre in Rotterdam: this was designed by Henry Moore and executed by two master masons.

The various ways of manipulating brick practised over the centuries are in themselves poor aids to the dating of early buildings. It can be said that walls of regular English bond are not earlier than about 1430 and walls of Flemish bond not earlier than about 1630. It can also be said that walls of haphazard bond rarely occurred after the seventeenth century. Bond takes us little further.

Brick sizes are useful for dating only in a rough and ready way compared with the shape of a building and with architectural features. Mediaeval brickmakers were not meticulous in constructing their moulds and the bricks of an old building may vary, even in a single course, by over half an inch in all dimensions. Sometimes bricks are present which were antique at the time of building: second-hand bricks have been made use of ever since the departure of the Romans.

Nevertheless, a study of Nathaniel Lloyd's table of brick measurements in his *History of English Brickwork* reveals broad changes in the thickness of bricks which are of some interest in considering the age of an old house. Up to the mid-fifteenth century most bricks were $1\frac{3}{4}$ inches thick. For the next two hundred years the measurement was 2 inches or a little over—

the bricks at Balls Park, Hertford, *c.* 1640, are between 2 and $2\frac{1}{4}$ inches. During the second half of the eighteenth century just over $2\frac{1}{4}$ inches was usual, though here and there, as explained in chapter 7, bricks are to be seen which are $3\frac{1}{4}$ inches thick and unusually long: these 'great bricks' almost certainly date to the period 1784–1803 during which it was possible to reduce the effect of the regulations of the brick tax, 1784–1850. Throughout the nineteenth century most bricks were 3 inches thick.

Brickwork of the nineteenth century 9

41
Spatial effect with brick was still important in the early nineteenth century. The stables at the Royal Hospital, Chelsea, designed by Sir John Soane, 1807

Brick had ceased to be a fashionable building material at the start of the nineteenth century; it was being associated with the great brick-built mills, factories and warehouses (which were splendid but to do with trade) and with cheap dwellings for the workers. A man of position now found that he did not care to live in a brick house, whether red or yellow, unless it was covered.

George III had something to do with the sweeping distaste for naked brick. Word spread fast, for instance, of his comment on visiting a fine brick house near Weymouth which its owner, Mr Morton Pitt, was accustomed to hear praised for its beauty. All the king said was, 'Brick, Mr Pitt? Brick?' (according to Arthur Oswals, *Country Houses of Dorset*). Mr. Pitt took the hint and went to the great expense of having his house faced in stone.

Apsley House (Number One, London) was a brick pile when completed in 1778, but in the 1820s the Duke of Wellington had it veneered with Bath stone, and at about this time Kingston Lacey in Dorset, a red brick house with stone dressings, had all trace of bricks hidden beneath a casing of Caen stone. There was no objection to building with brick, provided it did not show. Stucco was of course the cheapest means of hiding it. The bricks of Colen Campbell's Mereworth Castle and Lord Burlington's villa at Chiswick were rendered over at the time they were built.

The prejudice spread to the most ordinary artisans' terraces: they were stuccoed up to the level of the first floor, thereby making necessary recurrent repair work of the sort those who live in them must still carry out. Throughout the Regency period and for years beyond, stucco or half stucco seemed to

most people an obligatory covering for the brick house.

But the voice of reason began to be heard in the 1830s. In his book *Observations on Building and Brick-making*, 1834, the travel writer Robert Bakewell said

> gentlemen who construct houses for the purpose of residing in themselves cannot be charged with parsimony, and scarcely with economy, otherwise the artificial, temporary and expensive practice of daubing over their fronts with mortar (fashionably styled stucco) would not so generally prevail; and although it must be acknowledged that houses of this description look handsome for a while, yet, comparatively, in a short time they become mean and shabby in the extreme and require to be almost continually patched or repaired, and even when most perfect, we know it to be a deception, and intended to hide deformity and worthlessness; and the defects abovementioned can be only partially amended by incurring the enormous expense of periodical painting at short intervals. . . .

It is true that the underlying bricks were sometimes defective, thrown together in the weakest mortar. Why trouble about the quality of the brickwork when it was all to be covered over? Today, where a Regency building is being pulled down, it is often possible to inspect brickwork of a kind suggesting that, but for the rendering, the rain would have trickled through. For these reasons it is seldom worth a house owner's while to disinter brickwalls from their stucco. In northern Italy during the last twenty years many old buildings have had their stucco stripped away in order to reveal mellow brickwork, and to save maintenance. But too often the work has exposed poor bricklaying and individual bricks too soft to stand the shock. Interior brick walls of churches in Spain and Portugal which have recently had the original plaster removed seldom look well because of fragments left adhering.

42
London stock bricks were used for 174 Gothic churches in the new areas of London during the first half of the nineteenth century. St John's, Walham Green, a 'commissioners' church' of 1820

By about 1840 undisguised brick was returning to favour, and it was fashionable to denigrate stucco as dishonest. Perhaps the use of good honest brick for churches was influential: new aesthetic effects with brick had been found during the building since 1820 of the churches sanctioned by Parliament for the growing new districts of London. Brick, which had been forced on the church architects for reasons of economy, largely dictated their choice of the Gothic style. This could be realised decently in brick whereas the classical style called for a large quantity of expensive stone for cornices, pediments and other inherent features.

Uncovered brick was admired once more for every kind of domestic dwelling house by 1852. In that year Ruskin published *The Stones of Venice* and argued that wherever possible brick buildings, from cottages to churches and factories, should be adorned with such devices as string-courses of different coloured bricks—and preferably shiny ones. The idea was taken up with zest, as may be noted in most parts of Britain. Polychrome brickwork in working class houses in Reading, for example, is to be noted in many small streets on both sides of the River Kennet. Sir Arthur Blomfield, the church architect, built a church at Shoreditch with bricks that were red, yellow and blue. Having arrangements of these colours on a wall face appealed to the great novelist and poet Thomas Hardy, who as a young man worked in Blomfield's office: in 1863 Hardy won a competition for the best essay on the application of coloured bricks to modern architecture.

During the 1860s red bricks became even more fashionable than buff ones. Elaborate brick houses in Kensington such as No. 1 Palace Green by Philip Webb, 1868, and Lowther Lodge by Norman Shaw, 1870, helped to start a demand for the type of house that became known as 'Queen Anne Revival'. The name was rapidly accepted, presumably because of the red brick, the white sashes to the windows and the disappearance of plate glass in favour of small panes, but the irregularities of

43
The house in London which started a fashion known (on slender grounds) as the Queen Anne Revival—No. 1 Palace Green, Kensington, designed by Philip Webb in 1868. Among the characteristics of the type are even red brick, plenty of white paint and irregularity

44 *left*
Ecclesiastical work showing the strength of Ruskinian influence, especially in the banded decoration—Butterfield's Keble College Chapel, Oxford, 1870

45 *right*
Brick, having come into its own, rejoices in picturesque magnificence. Alfred Waterhouse's Prudential Building, Holborn, London, 1879

the designs (the opposite of symmetrical), the terracotta finials and the barge boards made a travesty of the architecture of Queen Anne's day.

It was now possible to buy machine-made bricks. The Victorians, admiring a precise finish, were delighted by the calculated uniformity of their colour and by a crisp regularity in shape which encouraged brickmakers to work with thin straight joints. These mortar joints could indeed be sufficiently thread-like to have no toning down effect on the impact of brick colour.

The brickmakers of the Midlands concentrated on turning out the sort of hot red brick which is well exemplified at Rushden, Northamptonshire. In the northern counties a tomato redness of (to us) uncomfortable intensity was presented, and still is,

by the smooth, hard-pressed bricks that resulted from exploitation of the coal measures. Mr Clifton-Taylor in his *Patterns of English Building* comments that these shales provide

> bricks which will unfortunately wear for ever without weathering at all. Accrington in central Lancashire has the dubious distinction of having made some of the most durable and visually disagreeable bricks in the country. Not for nothing is one of the two principal varieties known locally as Accrington bloods.

In the south many people were still happy enough to build with mellow stock bricks made by hand, though villas of a smooth red appeared even in rural parts.

Mortar in those days was not expected to stick the bricks together so much as to keep them at the right distance apart, and lime, not cement, was the active ingredient in the mix. Cement of various kinds had always been prepared for particular jobs, but the efficient Portland cement of today (it consists of limestone and clay calcined by burning at a high temperature) was hardly on the market for general use till around 1900. Even so, because of the transport and packaging factors, it remained more usual till after the First World War to make mortar with lime and sand than with cement and sand.* The difference between the two is obvious when a wall is taken down. Lime mortar does not cling, limpet-like, to the bricks: the ease with which, compared with a cement and sand mixture, it can be knocked off them is of importance to the trade in second-hand bricks.

Lime mortar was nevertheless an excellent, easily applied material, when carefully made—as it was in London for large buildings going up under the direction of an architect or a good builder; but too often it proved in lesser work a friable mixture

* The original north wing of Buckingham Palace was erected on foundation concrete made of lime and stones (and no sand). When the building had to be taken down, it was found that the stones, so far from setting into a mass, were still loose: D.H. Mahan, *Civil Engineering*, Fullarton, 1870.

46
A cartoon in *Punch*, 1875, during the height of the bad mortar era—Trollope's *The Way We Live Now* was published in that year

'The Way we Build now'

Indignant Houseowner (*he had heard it was so much cheaper, in the end, to buy your House*). 'WH' WHAT'S THE—WHAT AM I!—WHA'—WHAT DO YOU SUPPOSE IS THE MEANING OF THIS, MR. SCAMPLING!?'
Local Builder. 'T' TUT, TUT WELL, SIR, I 'SPECTS SOME ONE'S BEEN A-LEANIN' AGIN IT!!'

always on the point of dropping out. Apart from a builder's wish to scamp and put more pounds in his pocket, there was the real difficulty over the cost of transporting the right materials to an urban site.

The architect William Bardwell in his book *Healthy Homes and How to Make Them* gave these directions in 1860:

> Mortar, to be durable, should be composed of well-burnt limestone, and sharp, clean river-sand—one of the former to two of the latter. The Thames sand, taken from above the bridges, is a pure drift-sand . . . and suffers but little diminution in its bulk by washing. The lime that heats the most in slaking is the best, and slakes the quickest when properly watered. . . . The mortar should be mixed in a

> pug-mill, or well-beaten, so as to thoroughly incorporate the lime and sand. Care should be taken to have it used soft in warm weather, and rather stiff in cold weather, and to insist upon all bricks being dipped in a tub of water. With mortar so made with pure water, free from salts or clayey particles, and with good stock-bricks so treated, you will have a homogeneous mass, solid as a rock, which will increase in strength every year.

Instead of doing these things builders tended to use as little lime as possible, and sand that was far from clean and sharp. Where the house was not tall and the walls were thick and solid, it might not matter—and in terraces the houses helped to support one another. Most people who have driven new windows or doors in Victorian cottages, especially in towns, will speak of their surprise at finding that once the work was started the bricks could be just lifted from their positions.

Some builders resorted to lime kept until it had reached a state of chalk and then mixed it with dust. Dr T. Pridgin Teale reported in his *Dangers to Health* (1878) on houses in Leeds where dust from the roads and the contents of ash pits were ground up with 'a bare pretence of lime' to make mortar. Fires, he said, were lit against the thin, propped walls to encourage the mortar to set.

Summonses were common. A typical summons was taken out at Hammersmith Magistrate's Court in November 1878 by the district surveyor. The *Builder* reported the case:

> The district surveyor said that the defendant, instead of using mortar compounded of lime one part and sand three parts, a very inferior substitute had been used, composed mainly of vegetable soil, slightly charred by the process of burning bricks upon the ground. He had frequently remonstrated with him, and had complained that the good sand, with which the neighbourhood abounded, had been dug out and sent away instead of being

47
An increasingly familiar figure in Victorian times, —as illustrated in Kenny Meadows' *Heads of the People*, c. 1845

used in the buildings. . . . The magistrate ordered the demolition of certain portions of the buildings, allowing the district surveyor his costs.

Proved mortar offences seem to have been at their most numerous in the 1880s and were dealt with swiftly in the courts. When a suburban builder was fined £3 with £2 3s. costs, the magistrate remarked that if a man chose to use mud instead of mortar, he would just have to put up with the consequences. The *Brick and Tile Gazette* suggested in 1886 that it might be useful to grind up old building materials as a substitute for sand, the stuff to be mixed with plenty of good-quality lime. Mortar so made had been well spoken of in Bradford, Blackburn and Birmingham.

In the period 1860–70 the number of houses built in England and Scotland, mostly of brick, was 53,300. During the next decade the number rose to 80,000. The use of brick in much poor-quality terrace housing did not help the material's reputation. But it easily survived disparaging comments. By the 1880s unadorned brick, still red if possible, was being taken very seriously for prestige architecture.

John Randall, historian of the Shropshire clay industry, had written in 1877: 'People have been so long accustomed to see brickwork used for inferior houses, and stone for buildings of greater pretensions, that until recently English bricks have scarcely had justice done to them.' Nature's finished material (that is, stone) might be deemed more suitable for churches, he said, but artificial stone (brick) 'fashioned into shape by man, was quite as appropriate for a dwelling in which gathered the highest social sanctities.'

Towards the end of the nineteenth century the existence of railway transport made it easier and cheaper to build with brick than stone. This has meant a certain loss of regional characteristics, for builders followed stock designs, and brick dwellings everywhere looked alike in shape if not necessarily in colour.

10 Victorian brickyards

Nineteenth-century mapmakers who plotted brickyards near London as landmarks often overlooked the tendency of these to move on. In most country yards the bricks were fired in sturdy kilns, but the makers of London's yellow stock brick had no kilns to tie them to a particular site. Their apparatus was simple and portable. A few years after moving on, a difference in the level of a field might be the only trace of where bricks had been made. A good brickmaker (in a clay district) could set up his horse-operated pugmill, shape the bricks and burn them in clamps almost wherever he considered his product was needed. Since the clay could not be moulded in frosty weather, brickyards operated only in the season from April to October.

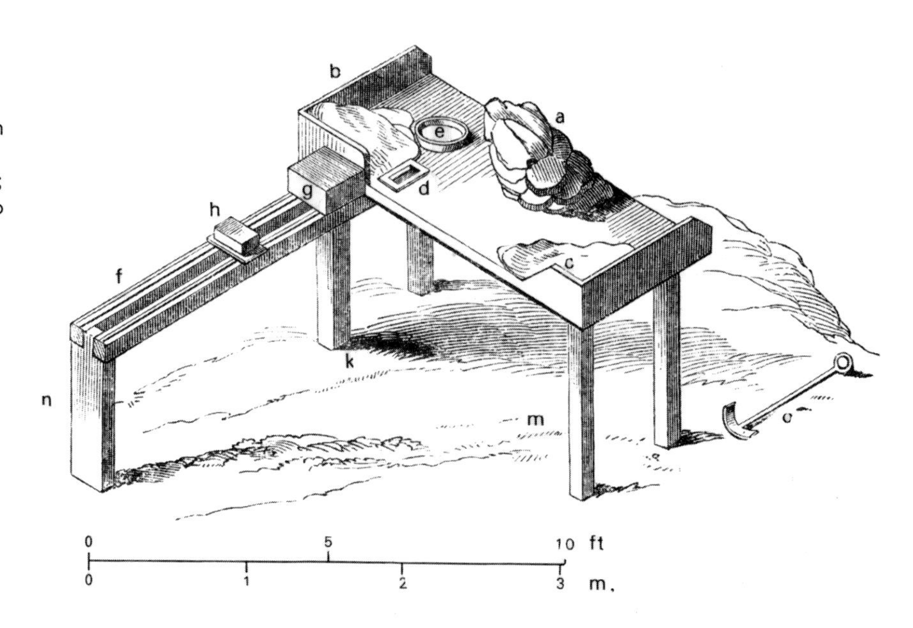

48
Moulding table shown in Edward Dobson's *Treatise on the Manufacture of Bricks*, 1850, where the items are thus identified: *a*, lump of ground earth from the pugmill; *b*, moulder's sand; *c*, clot-moulder's sand; *d*, bottom of the mould termed the stock board; e, water tub; *f*, the page, formed of two rods, on which the new bricks with their pallets are made to slide away from the moulder with facility; *g*, pallets in their proper position for use; *h*, newly-made bricks ready for the taking-off boy; *k*, the moulder's place; *o*, the cuckhold, a concave shovel used for cutting off the ground earth as it is ejected from the pugmill.

49
Cartoon on jerry-builders by George Cruikshank, 1829, showing the farmland of Hampstead being devastated by bricks bursting from their clamps

The idea of running a brickfield appealed to many men because of the simplicity of the process, and tiny fields using shallow deposits of clay abounded on the fringes of London. They were operated just ahead of the advancing builders who in due course consumed them. By the middle of the century 500 million bricks were being made every year within a five-mile radius of London Bridge; and, although this was three times the quantity made in 1820, the pugmill remained at that time the only

and begrimed by the smoke of kilns and domestic coal fires. Like the so-called march of intellect, the march of mechanical invention was a favourite theme for satirists in the 1820s

mechanical invention to be employed. Although brickmaking was healthier than much Victorian industry, and not listed in *Dangerous Trades*, results were achieved by labour that proved too much for the physique of many employed in this way.

Brickyards in the London area generally opened at 4.30 a.m., according to the *Fifth Report, Children's Employment Commission* of 1866, and till 8 p.m. most were scenes of continuous toil by women and children as well as men. A foreman told inspectors: 'One half-hour at 3 p.m. is the only real meal time they have in the day—their meals are brought to them and they swallow them down as fast as they can.' Because brickmaking was seasonal and payment made per 1,000 bricks, there was the urge to earn all the money possible while the work lasted. But the men commonly preferred buying drink to saving up for the winter. It was said at one yard: 'They are too fond of bringing children to work . . . they spend most of their earnings on drink and then look to the children's earnings to pay for rent and food and clothing for the family.'

There was a beer shop in every brickfield, according to the Rev James Dennett, a missionary. He said in his evidence: 'Drunkenness is the curse of the working man in every trade in this country, but it seems tenfold intensified in that of brickmaking.'

One foreman said the hours at his brickfield had previously been 3 a.m. to 8 p.m., but now they had made them 5 a.m. to 7 p.m. and it was found that the work done was no less. The horse was put on the pugmill at 6 a.m. and taken off at 6 p.m. Two suggestions for brickfield operation were offered by the foreman: that Saturday should be a proper half holiday and that wages should not be paid in public houses. 'They don't leave off in most fields till 3 or 4 p.m. on a Saturday. Then they all have to go to the public house to be paid, even the little boys and girls, who are thus, you might say, taught drunkenness in their childhood. . . . Worst of all is the practice, though unknown to employers, I know to exist: the publican allows the

foreman a percentage upon all the money which he pays out in wages at the public house.'

A boy of nine who was questioned said it was his first summer in the brickfield. 'I am a barrow loader. I never work later than a quarter past seven, and I'm a good 'un at getting up— I always get to work by 5 a.m. I am very tired when the day's work is over.' A common task for young girls was carrying to the moulding tables lumps of clay weighing up to 24 lb. 'In the

50 *left*
Carrying clay from pugmill to moulders, 1871

early spring after a frosty night,' it was stated, 'the damp clay strikes very cold to the chest and stomack.'

Brickfield children went to school in the winter months and were supposed to attend Sunday school during the season. A headmaster commented: 'What the children learn in the winter they have almost entirely forgot in the summer. . . . Two months are needed each winter to make a revision of their previous knowledge.' The effects on their conduct included 'filthy and blasphemous expressions, sullen demeanor, habits of untidiness and dispositions of cruelty'. In 1871, the reformer George Smith published *The Cry of the Children from the Brickyards of England*; it aroused the interest of the seventh Earl of Shaftesbury and in the same year came the Factories Act (Brick and Tile Yards) Extension providing for child and female labour to be regulated. According to a Press report, 'on the first day of 1872, 10,000 young children were sent from the brickshed to their homes and to school.'

The manufacture of satisfactory bricks was hardly suggested by the appearance of London's small brickyards. A feature of each was a large heap of ashes and other house rubbish. This provided both an ingredient for the bricks themselves and a fuel for burning them in clamps, and the smells to which it gave rise in hot weather led to several successful actions by neighbours. The great nineteenth-century authority on ceramics, Professor Hermann Seger of Germany, was mystified when he visited London in 1872 and went to see some of the brickfields. 'How', he asked, 'can bricks be made, suitable for house fronts, without a plant, dryer or kiln, especially since the bricks are to be used not for rural building, but for structures in the English capital?' The answer is that the brickmaking material was more carefully mixed than he realised and that the ingredients, surface clay, ashes and chalk, were well suited for kilnless firing in open clamps. What resulted, in various grades, were London's famous yellow stock bricks.

Despite the primitive-looking means of making, these bricks

of London and elsewhere in Britain were superior to much German brick (Seger reported an uncharacteristic lack of care in German yards). They were also better than American products. Mid-nineteenth-century American bricks were largely 'slush stock', according to Charles T. Davis in his *Treatise on the Manufacture of Brick*, 1895. They were made of under-mixed material cut by wires in a large shallow box without being first compressed either by hand or machine. After burning (or

51 *left*
Brickmaking in London as shown in *The Boy's Book of Trades*, 1871

52 *right*
Brickmaking in Worcestershire photographed in 1902—from *Britain at Work* (Cassell)

under-burning), the rather cracked bricks were 'light, very open or porous, therefore absorbed water readily. . . .' The brickmaker was concerned with little else but turning out the requisite shape. Towards the end of the century many house walls in Chicago were falling apart, and everyone was glad that by then there were techniques and machines capable of mass-producing bricks efficiently.

In London stocks, the chalk, which was added in the form of a paste, reduced a tendency to crack while drying and formed, during firing, a silicate of lime and alumina. It also gave the bricks their colour, which would otherwise have been red. The ashes, containing a proportion of coal dust, helped the bricks to reach a high temperature in the clamp. Each was a sort of fireball, in the phrase of Edward Dobson whose *Treatise on the Manufacture of Bricks and Tiles*, 1850,* gives a detailed account of making bricks in London, a procedure which varied little in essentials elsewhere.

For the work of actual brickmaking to begin, the moulding stool was provided with two heaps of dry sand, a tub of water, a stock-board and brick mould, and three sets of pallets. When a supply of tempered clay had been placed on the stool by some-

* A facsimile edition of this landmark in brickmaking history has been edited by Dr Francis Celoria, Keele University, and published in 1973 by the George Street Press, Stafford.

53
Bricks being fired in clamps in northern France

one called a feeder, the clot-moulder (generally a woman) sprinkled the stool with dry sand and, taking some clay, kneaded it roughly into the shape of a brick and passed it to the moulder on her left hand. The moulder then dashed it with force into his previously sanded mould. With his strike, well wetted in the tub of water, he removed the unnecessary clay, throwing it back to the clot-moulder to be re-moulded.

He then turned out his raw brick onto a pallet, slid it along to the taking-off boy and re-sanded his mould in preparation for the next brick. A moulder, helped by the feeder, clot-moulder, taking-off boy and two people to wheel away and set the bricks to dry, would make about 3,500 bricks a day between 6 a.m. and 6 p.m.

The raw bricks were stacked to dry in rows eight bricks high, known as hacks, and protected from rain, frost and heat by a covering of straw. When half dry they were scintled (placed herringbone-fashion) to allow the air to pass freely between them.

When quite dry, at the end of three to six weeks, the bricks would be set for burning in vast heaps (or clamps) of 100,000 or more and encased with burnt bricks. Fuel in the form of cinders was skilfully distributed in layers between the courses of bricks, with a few especially thick layers at the bottom. To light the clamp, several flues were left and filled with faggots. These, when lit from the outside, soon set fire to the adjacent cinders. As soon as the clamp was fairly alight, the mouths of the flues were stopped and the clamp was allowed to burn itself out, a process taking from three to six weeks.

The bricks at the outside of the clamp were underburnt; they were called burnovers and were laid aside for re-burning in the next clamp. Bricks that were only a little underburnt were called place bricks and sold off cheaply as merely suitable for inside work. Bricks near the live holes which partially melted were called clinkers and sold by the cartload for rockeries in gardens. There was much sorting out of the bricks when a clamp had been burned. Here are the various London grades as listed by Dobson in 1850:

Cutters	the softest, used for gauged arches and other rubbed work.
Malms	the best building bricks, only used in the best descriptions of brickwork; colour yellow.
Seconds	sorted from the best qualities, much used for the fronts of buildings of a superior class.
Paviours	excellent building bricks, being sound, hard, well-shaped and of good colour.
Pickings	good bricks but soft and inferior to the best paviours.
Rough Paviours	the roughest pickings from the paviours.
Washed stocks	the bricks commonly used for ordinary brickwork and the worst description of malms.
Grey stocks	good bricks, but not of irregular colour and not suited for face work.
Rough stocks	very rough as regards shape and colour, not suited for good work although hard and sound.

Grizzles	somewhat tender and only fit for inside work.
Place bricks	only fit for common purposes, should not be used for permanent erections.
Shuffs	unsound and full of shakes.
Burrs *or* Clinkers	only used for making artificial rockwork for cascades or gardens.

The permanent brickfields of north Kent, especially those by the estuary around Sittingbourne, were another source of clamp-burnt yellow bricks—they are made there still. Frank G. Willmott describes in *Bricks and Brickies*, 1972, how the fuel was acquired. The sailing barges taking bricks up the Thames to London used to return with loads of dustbin emptyings from the wharves of Vauxhall, Chelsea and Putney. Swarms of flies accompanied the barges, and crews had to beware of spontaneous combustion and carbon dioxide fumes from the coke. The rough stuff, as they called it, was tipped into large mounds and left for a year or so for the vegetable matter to rot away; then it was sifted and graded.

Anything of value found in rough stuff was considered moulders' perks. Sometimes trinkets and brooches turned up

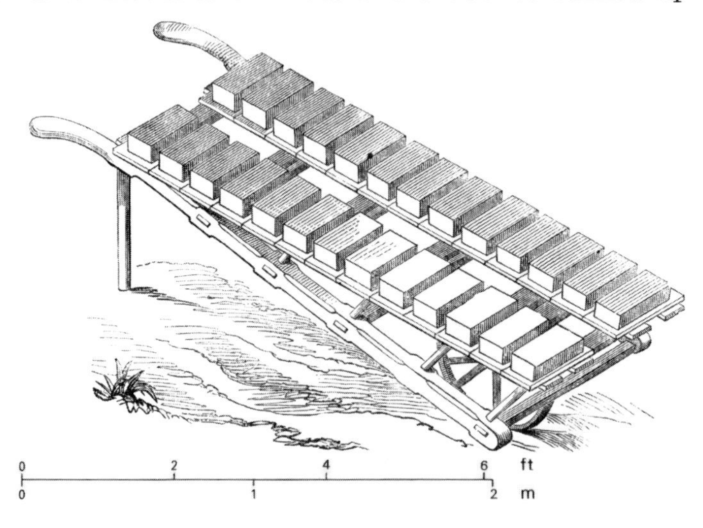

54
Barrow for twenty-six green bricks illustrated by Dobson

in refuse from Chelsea where the big houses were. Fragments of china and glass were dumped at Lower Hadstow at a place which became known as Glass Bottle Beach. The fine ash was mixed with the clay and the bigger lumps laid under the clamp to burn the bricks.

This use of rough stuff continued until 1965, when they found it was not burning satisfactorily: householders were using the smokeless fuels which produced an ash practically useless for brickmaking. However, brick men have since found a supply of similar fuel in the dust obtainable from coal washeries.

At other brickyards in Victorian England the firing was generally done in Scotch kilns fuelled by coal or wood. Everywhere the moulder, dashing lumps of clay into his wooden mould, was the key worker at a muddy place of work employing many hands. He chose and paid his assistants, who were often members of his family, and with their help would turn out over 2,000 bricks a day. J.H. White, in a report on brickmaking in the Midlands, wrote in 1864 of boys hurrying to and fro with mounds of clay on their heads and of mud-splashed, bare-kneed females singing coarse songs as they tempered clay with their feet or laboured at the turning and shifting of bricks in the making.

He gave an account of women and children drawing a kiln at Brierley Hill. They were in a line, tossing the warm bricks to one another, two at a time, to a cart on the road.

> A small girl of 12, forming one of the line, struck me by the earnest way in which she was doing her share of a work which is certainly heavy for a child, as a slight calculation shows. The kiln, containing 17,000 bricks of 7¼ lbs each, was to be emptied by ten persons in a day and a half: i.e. this girl had to catch and toss on to her neighbour in a day of only the usual length a weight of more than 36 tons and in so doing make 11,333 complete turns of her body. When called down by me she was panting.

Victorian inventions 11

The second half of the nineteenth century saw the birth of numerous machines for the brick industry; versions of some are still in use little changed today. Inventiveness was encouraged by a demand for bricks so great it could not have been met by traditional methods alone, with the labour available.

55
An agreement of 1883 to mould bricks at the rate of 4s. 8d. per 1,000

But brickmaking machines had to be introduced slowly. In the north of England, especially, they tended to be sabotaged as job-stealers; there were cases of workers smashing them at night and of much damage being done by the simple trick of dropping a steel spanner into the grinding mechanism. There was a flagrant case at Manchester in 1861 when brickmakers Renshaw and Atkins had their steam engines maliciously blown up. Decades were to pass before such Luddite activity ceased entirely.

Even in the regular yards where machinery had been accepted, brickmaking remained arduous and dirty; it remained, too, till Edwardian times, a largely seasonal activity with no work to be done in the winter. 'But life was much simpler then', an old brickmaker told *Redland News* in 1972, 'and we managed to have a lot of fun.' He had been a member of a gang of six making stock bricks in north Kent and vividly remembered the competition between the gangs to make the largest number of bricks in the yard. The target was a million bricks in a season. Achieving this number carried prestige and earned the gang the bonus of a new pair of boots. 'Every man jack in the gang had to work to split second timing for as much as ten hours a day,' the old man said, 'so you can tell it wasn't just the bonus we were after.'

But it was the spread of the newly-invented machines much

BRICKMAKER'S AGREEMENT.

ARTICLES OF AGREEMENT FOR SERVICE, made this 31st day of March 1883, **Between** Thomas Rowe Brickmaker and Moulder, of the Parish of Tottenham, Middlesex, and Thomas Plowman jr & Mark Plowman trading as THOMAS PLOWMAN, Brickmaker, of the same place. The said Thomas Rowe hereby agrees for the consideration hereafter expressed, to serve the said Thomas Plowman in his capacity as a Moulder of Bricks for the Brickmaking Season of 1883, at his Brickfields at Edmonton the commencement and termination of such Season to be regulated and determined by the said Thomas Plowman.

The said Thomas Rowe hereby agrees to work the full time allowed for work by the "Factory Act," or the "Workshops' Act," whichever the Field may be classed under, whenever the weather shall permit him to do so, and to execute the work in a good and workmanlike manner, to the satisfaction of the said Thomas Plowman; to leave off Moulding at any time rendered necessary by the weather, and to begin Moulding again when the weather is fit for that purpose; and to thatch and unthatch his Bricks when required by the said Thomas Plowman, or his authorized agent; and not to neglect or delay the work in any way: and the said Thomas Plowman hereby agrees to employ the said Thomas Rowe for the Brickmaking Season of 1883, and to pay him at the rate of Four Shillings and eleven Pence per Thousand if the earth is pugged by horse, or four Shillings and eight Pence per Thousand if the earth is pugged on to the table, for all Bricks properly made; and a further sum of seven Pence per Thousand, at the end of the Season; no such further sum to be payable if he neglect or desert his work, or is discharged for a just and sufficient reason, whereby the Season is not completed.

As Witness our hands

Thomas Rowe

Thomas Plowman

Witness Robert Burton

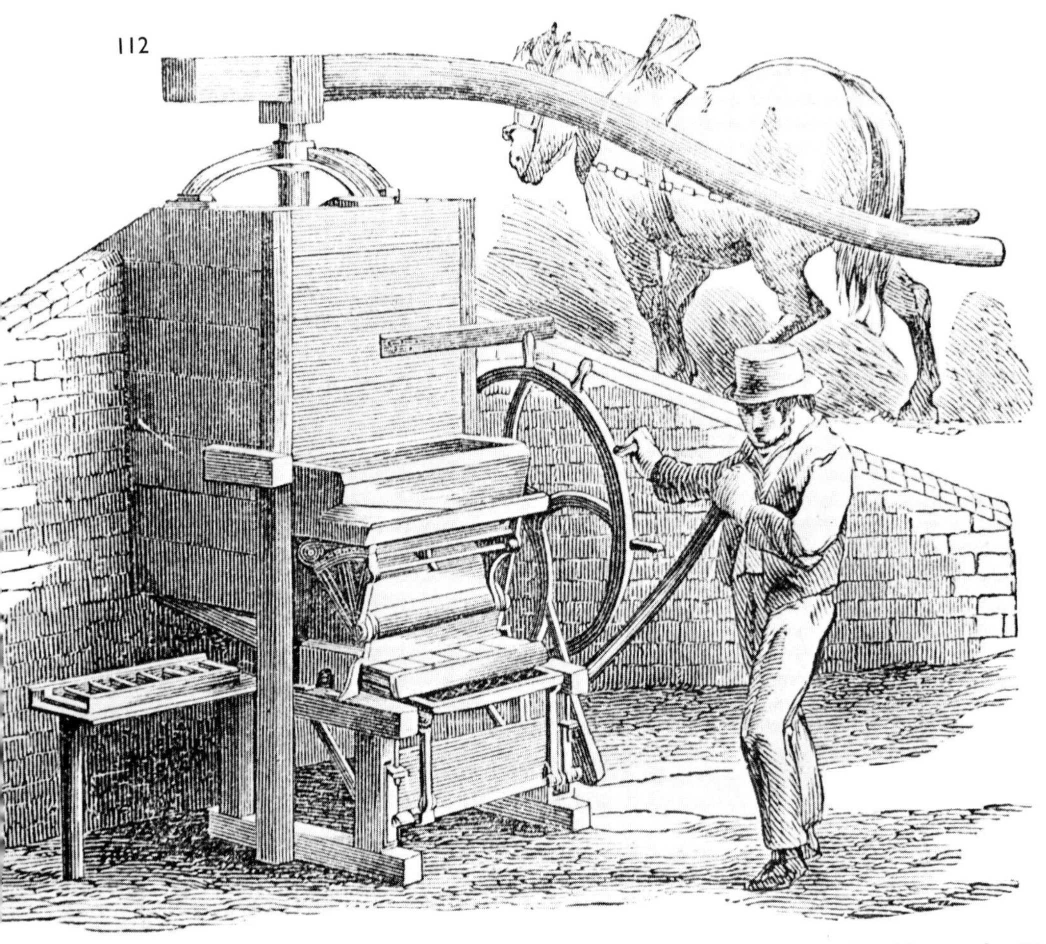

56
Hall's horse-assisted machine of 1846 for making bricks by the moulding process. It was said to be possible to make with it 10,000 bricks in a day

more than hard work which allowed Britain's production of bricks to double in the period 1850 to 1900. The one great human benefit eventually brought about by mechanisation—and by better buildings—was that many firms were able to work their clay all the year round and provide steady employment.

The ingenuity of the nineteenth-century inventors is shown clearly in the illustrations. The first brick-making machines imitated hand-moulded bricks: a quantity of soft clay was dropped into moulds and then compressed by passing the moulds under the cylinder of a vertical pugmill. Early machines of this type were Lyle and Stainford's pugmill-brickmaker of 1825, designed to make fifteen bricks at once, and the Jones

57
Hand-operated machine by Bulmer and Sharp, 1861, said to be capable of turning out 5,000 wire-cut bricks in ten hours

machine of 1835 which had moulds fixed on a rotating table.

An American machine by Nathaniel Adams was an improvement. It was driven, as usual, by a horse. When an attempt was made in 1840 to drive it by steam, a disgruntled mob prevented the steam engine from being started and actually drove Nathaniel Adams and his family from their house in Philadelphia; they had to stay away for a fortnight. The Barnhart steam-driven machine for winning clay, developed in the 1880s and first installed at a works on the banks of the Ohio River, was welcomed by the men—although with two operators it did the work of eight—because digging by hand was made dangerous by landslides.

The output of a steam brick machine could be 1,000 bricks an hour compared with about 3,000 turned out in a full day by a hand-making gang of six; but the direct expenses of working were of course higher with machines and constituted one good reason why the small works, the most numerous kind, were slow to invest in them. Although powered pugmills were considered a necessity by the end of the nineteenth century, the small yards were still winning clay by hand, and tempering and

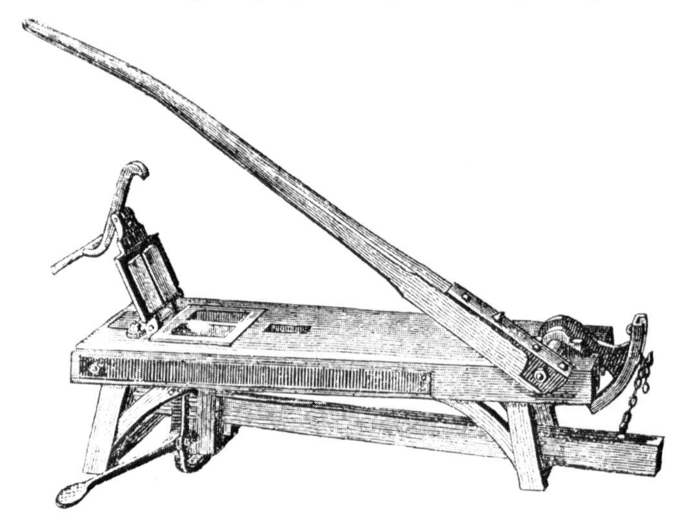

58
A hand-operated double lever press of about 1850 designed to deal with hard clays or to make dense bricks. The stocks were raised by means of a foot pedal, thereby easing out the bricks. It was claimed that a man and two helpers could produce about 300 bricks an hour with this press

moulding it by hand. On the Continent, the far from small Wienerberger works near Vienna used immigrant labour so cheaply that in the 1870s it was turning out by hand over 200 million bricks a year. Seger reported seeing 1,000 individual moulding tables with a man and a woman at each—unlike in Britain, the woman did the moulding. In America, even the Clippert Brick Company of Detroit—according to a study of its records by Joseph E. Zias—had no brickmaking machinery until 1909.

The Victorian taste was for smooth exact bricks, but now enough people wanted the look of the hand-made article for some makers to try to imitate it when producing soft-mud

59
Early nineteenth-century wire-cut machine to be operated by a horse

bricks mechanically. One of the earliest inventions for doing this, after a fashion, was patented by R.A. Norris in 1899. His very first machine was in full work for thirty-five years following its installation.

The Norris machine gave the clay a light pugging and forced it below a descending plunger which filled a mould (or series of moulds) at a stroke. On the plunger rising, the mould was pushed forward, the surplus paste struck off and the mould bumped to loosen the brick within and turn it onto a pallet board. Each mould made three bricks at a time only, thus giving the operator plenty of opportunity to clean, sand over and replace the moulds. The machine could be worked by a horse. When driven by a three horse-power engine it would make about 8,000 bricks a day, if all went well. Just after the turn of the century a much more powerful machine, with an output of 20,000 bricks a day, was introduced by T.C. Fawcett of Leeds.

The wire-cut process originated in England and spread all over the world. In this the clay is forced through a die (or mouthpiece) so that it issues in the form of a shiny column from which bricks are cut off by wires. The possibilities of this method were first demonstrated by William Irving in 1841, but the first fully operational wire-cut brick machine was designed by Richard Bennett in 1879: the complete plant was in continuous use for fifty-two years. Extrusion was done by means of a piston which pushed the mass of clay between knives placed at right angles to each other. With this method the process could never of course be continuous, for once the so-called stupid had reached the end of its stroke, it had to be withdrawn and the machine re-filled with clay.

The next invention for the wire-cutting method was the screw extruder or auger machine. In Britain this is now practically the only device used for wire-cut bricks. In its simplest form, the machine consists of a cylindrical barrel within which rotates a close-fitting helix (like a corkscrew) mounted on a central shaft. The clay mix, having been fed in at an opening at the top of one end, is forced along the cylinder by the flights of the helix and consolidated in the flights, in a spacer section, and in the constraint of the eventual die. Today modifications in wire-cutting machinery may include a device for de-airing to give a denser brick, a sand squirt for changing the texture of the emerging column or a row of spikes for making it rugged. Quite often core pieces are fixed in the die to make tunnels along the length of the column and produce light, perforated bricks.

Early wire-cut bricks were not nearly accurate enough in form to be accepted by Victorian architects as facing bricks, and in

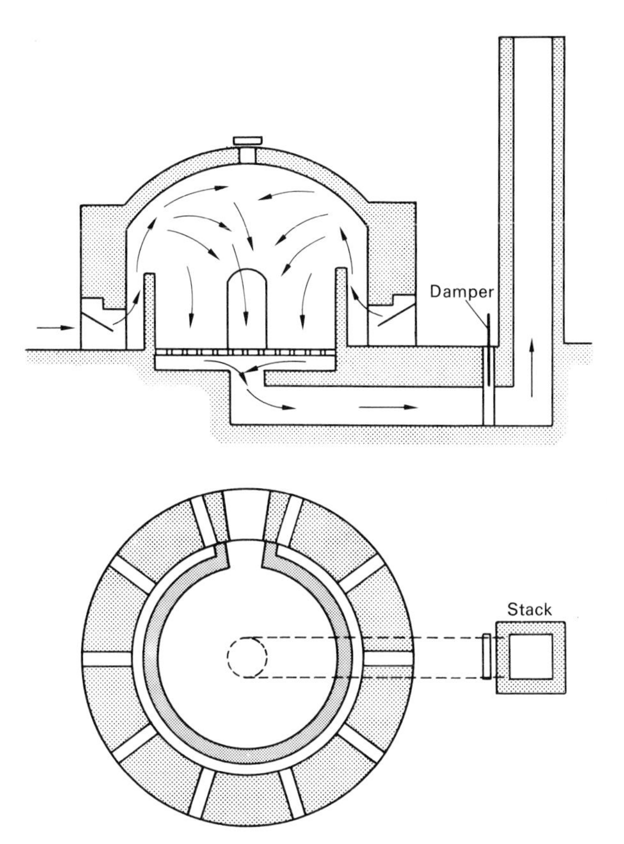

60
Diagram of a circular, down-draught, intermittent kiln. Hot gases from the fire holes round the circumference are drawn up and then down through the bricks on their way to the mouth of the chimney, which is shown on the right.

consequence they were customarily re-pressed. Machines for re-pressing date from the 1870s—an early one was by Bennett and Sayer. When effectively performed, the operation could have the effect of forming a dense brick with sharp edges. It was found that before the more plastic bricks could be pressed to advantage, these should be dried to an intermediate stage known as leather-hard.

The most important invention of the nineteenth century, perhaps in the whole history of brickmaking, was the invention in 1859 of the Hoffmann kiln in which the fire need never go out. It brought a great saving in fuel and allowed many more bricks to be produced.

Before 1859 all bricks were either burnt in open clamps or in single-chamber kilns called intermittent because the fire went out after each burning of bricks. And having particular uses, such as securing special qualities and colour, some of these kilns still exist in the yards of small firms. The simplest kind, cheap to erect, is a brick-built rectangular box with an open top and no chimney. The fires burn at fire holes along the two longer sides and the hot gases rise among the bricks, finally escaping at the top. This is called an updraught kiln.

61
A round Victorian-type kiln which is in fact a Hoffmann Continuous kiln and contains a ring of separate chambers linked by trace holes through which the fire passes. This kiln was still being used in 1973 in Suffolk

The other kind, the downdraught kiln, allows strict control of the heat and produces bricks of consistent quality. Downdraught kilns are usually circular but work as well when rectangular. They differ from updraught kilns in that the gases, having risen to the top, are deflected downwards among the bricks, and then pass through the floor to an underground flue leading to a chimney. To give an example of fuel needs in terms of coal, downdraught kilns may burn 8 to 15 cwt for 1,000 bricks, while only $2\frac{1}{2}$ to 4 cwt are needed to burn this number equally well in a Hoffmann continuous kiln.

In the invention of Friedrich Hoffmann, an Austrian, a number of chambers for burning bricks are placed side by side to form an endless series. The gases from one chamber pass into the next, then on to another, and so on until they become too cool to be of further use and pass up a chimney high enough to have created a good draught. When the bricks in one chamber are done the fire moves on. This is arranged by having a partition of newspaper in the openings (trace holes) between the chambers. The paper is first scorched and then ignited, by which time the gases in the kiln due to fire are exactly ready to ignite also. At the top of each chamber there is a small bung-hole through which coal dust is dropped as necessary to give the required temperature. Visitors enjoy looking down these holes during a walk round the flat roof of a Hoffmann kiln.

The small amount of fuel needed, compared with the requirements of the other kilns, is explained by the fact that the

heated air either remains in or is circulated into the temporarily fireless chambers—where it may help to finish the drying of green bricks which ideally are stacked there. To take full advantage of the Hoffmann kiln, a brickmaker should of course have a regular output of green bricks. Some brick firms can say with pride that their fires have not gone out in fifty years and more.

Another continuous kiln is the type in which the bricks move slowly on steel trucks through tunnels of flame. The tunnel kiln was first thought of in Britain in 1845, but until the last decade of the nineteenth century it could not be made to work properly. Today it is found convenient for firing by gas or oil and is used increasingly instead of the Hoffmann kiln.

62
Steam-powered machine by Bulmer and Sharp, 1861, said to produce 25,000 wire-cut bricks from unprepared clay in ten hours. In 1861 a firm of brickmakers at Manchester had their steam engine maliciously blown up as a job-stealer

63
Clayton's machine of 1860, with young man feeding in the clay

The advent of these efficient continuous kilns led to developments throughout Europe in brickmaking with the harder clays. The tough colliery shales would be ground to powder, mixed to plasticity with water and then shaped in power-driven clot moulds. This stiff plastic process, as it was called, produced strong bricks and was important to the coalfield areas of Lancashire and Yorkshire; bricks were needed there in quantity for factories and for housing—and also for works to do with the expanding railway systems. All kinds of mechanical inventions came with the discovery that coal-measure shales could be turned into bricks: things like chain haulage, dry pans with perforated bases, bucket elevators, rotary and piano-wire screens.

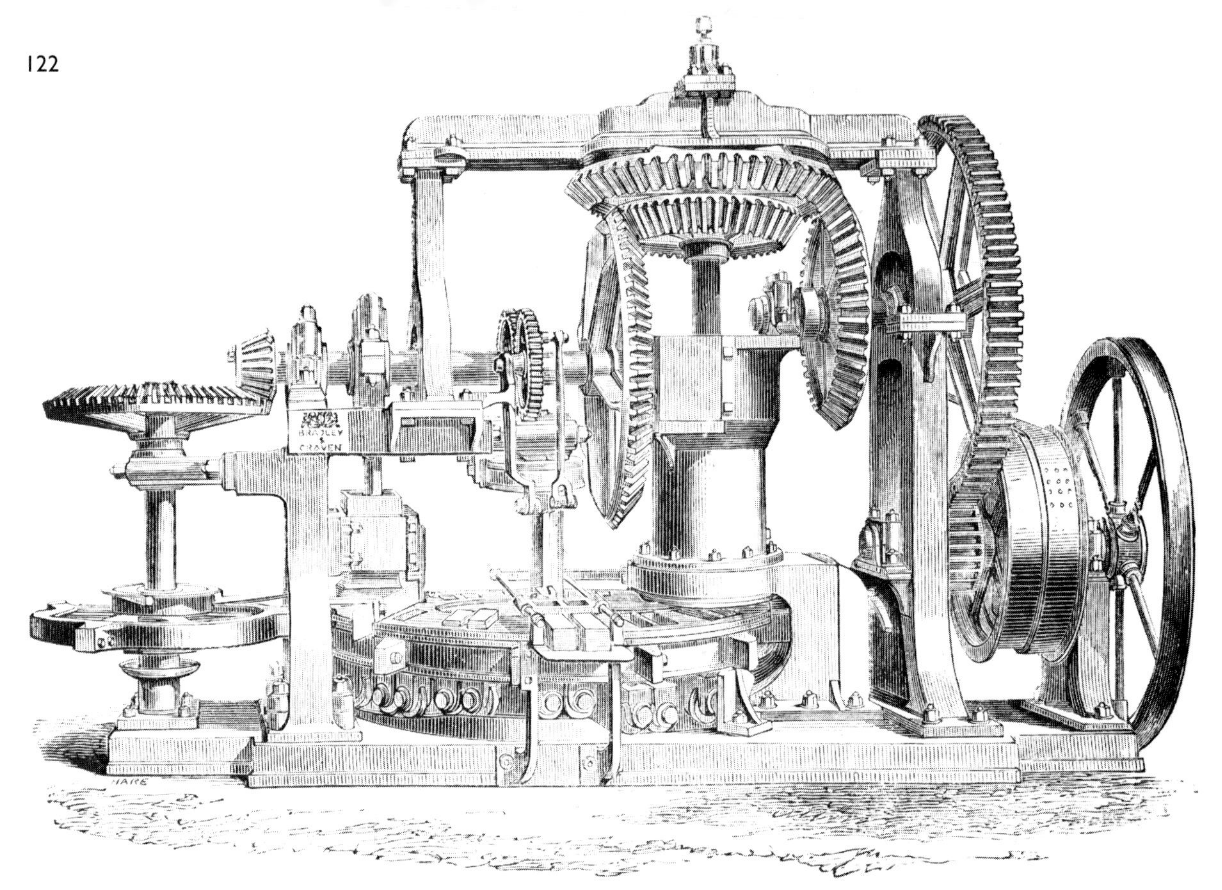

64
Bradley and Craven's machine for moulding bricks, 1860

In the late 1860s it was found possible, at Accrington, to make bricks by submitting shale dust to great pressure without the addition of water. This was the beginning of the development of the famous semi-dry process of pressing bricks which was later used in the fletton process and by which nearly half Britain's bricks are now made. How it works, and how it was exploited in the early 1880s at Peterborough, is described in the chapter on flettons.

A mid-nineteenth-century invention of much historical interest today—in view of its application in various parts of the world—is the extruded, geometrically shaped, hollow brick invented in 1849 by Henry Roberts. An illustration of it is shown beside a modern Swiss example employing the same principle.

Roberts's bricks, or anyway the idea behind them, was written of with enthusiasm by the architect William Bardwell in his *Healthy Homes* of 1860 and with no less enthusiasm by the present-day authority on bricks, Basil Butterworth, in a paper of 1956 published by the British Ceramic Society. The object of the bricks was to secure a bond without vertical joints passing through the wall and without headers. In their original form

65 *left*
A hollow brick invented by Henry Roberts in 1849 which had a brief vogue after being used in model houses shown at the Great Exhibition of 1851

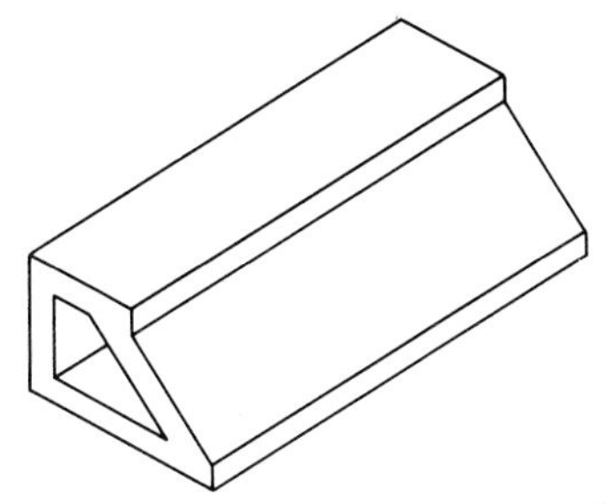

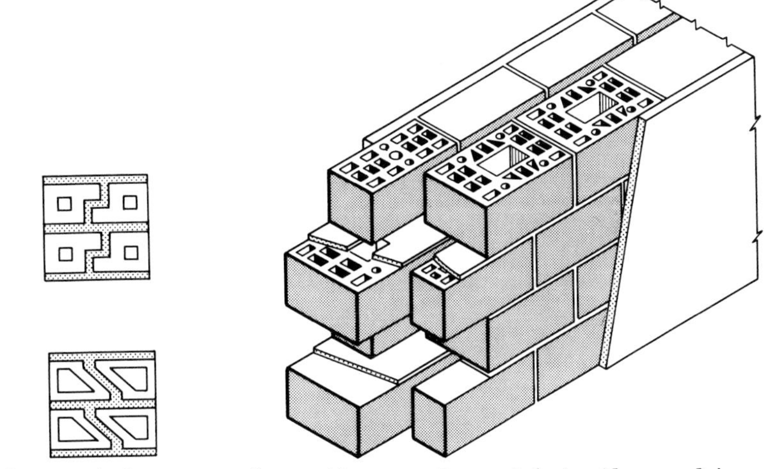

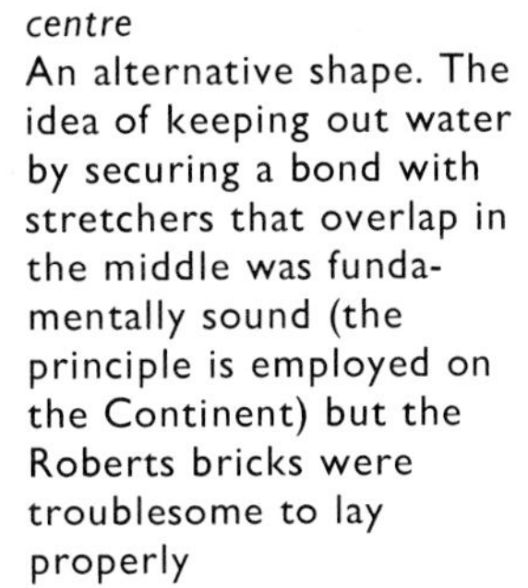

centre
An alternative shape. The idea of keeping out water by securing a bond with stretchers that overlap in the middle was fundamentally sound (the principle is employed on the Continent) but the Roberts bricks were troublesome to lay properly

right
Achieving the same result in Switzerland today by using two bricks of rectangular section but different widths

they had a certain vogue for a time and went into the making of some model houses for labourers shown at the Great Exhibition of 1851; but bricklayers found them troublesome to work with and the coming of the cavity wall made them at length seem dead ducks. In 1951 A.M. Foyle made these comments in *Brickbuilder*:

> Two defects probably caused Roberts's brick to fall into disuse: the number of specials required and the difficulty in handling the bricks that had to be laid with the wider of the two bed faces upwards. The principle of bonding stretcher bricks by making them overlap in the middle of the wall is, however, fundamentally sound, and it would have been but a step from Roberts's patent to achieving the same result by using two bricks of rectangular

> section but different widths. This step was never taken, however, though Roberts himself, in his original patent, suggested a transitional form. . . . To see the development that Britain might have made in 1850–70, we have to turn to Switzerland in 1930–50.

There is also the comment that, without the use of technically unwanted half-bricks, the bond presented to view could only be stretcher bond. This is no drawback in countries such as Switzerland where the rendering of outside walls is traditional and accepted, but wide use of the Roberts hollow brick in England would have made duller our heritage of Victorian brickwork.

66
Multiple mould, wooden but with iron facings, for use with a brick-moulding machine

12 Brickmaking for canals, railways and roads

Bricks have had an important part in making the routes by which they are now distributed, and have contributed to road building hardly less than to building canals and railways. The canals had especially the double association with bricks, for no sooner were they built and filled with water than navvy-like brickmakers, unpopular with farmers, tended to settle here and there on the brinks.

It was largely thanks to the readily-found brickearth that many canals and their locks and bridges came to be built; for carriage of building stone from a distance, wherever necessary, would often have made the cost prohibitive. A visitor watching work in progress on the Bridgewater Canal, which was opened in 1761 to transport coal from the Duke of Bridgewater's collieries at Worsley, wrote in the *Gentlemen's Magazine* that 'the duke, like a good chemist, has made the refuse of one work construct the material parts of another.'

He was referring to the usefulness in building Britain's first canal of the spoils of digging the cut. Nearly all the time, it seemed, clay was being brought up and employed both for the banks, to render them watertight, and for making the bricks which were vital to the duke's works (in the end forty-six miles of underground canals at Worsley were lined with brickwork for most of the way).

Occasionally good hard rubble stone appeared. In the early days of canal building, rubble stone, presenting itself as almost ready-made building units, was the material an engineer liked best to see turning up in excavations. But the quality of stone varies greatly from district to district. The builders of the Kennet and Avon Canal used defective Bath stone which was

67
The Duke of Bridgewater, the great canal builder, demonstrating his improved method of navigation. Among his achievements were forty-six miles of underground canals at Worsley, Lancashire, lined in brick for most of the way

responsible for troubles, dating from 1802, with locks, bridges and aqueducts.

In his book *The Kennet and Avon Canal*, 1968, Kenneth Clew explains that 'Bath stone is an excellent building material if certain precautions are taken before use, namely to ensure that the stone is properly seasoned by leaving it in the same position that it occupied before quarrying, for some length of time.' But the contractors did not do this, and they were careless about the selection of stone, with the result that some of the masonry work was a disaster.

The Kennet and Avon Company opened their own regular quarry—in fact two of them—but troubles with stone continued. The lock at Bradford was beginning to crumble and so were the aqueducts at Avoncliff, Biss and Semington. John Rennie, the canal's engineer, was in favour of using bricks instead of stone and in March 1803 wrote as follows, Mr Clew relates, to the committee:

> Seeing the great Loss that the Company have sustained and the great detention which the Works have experienced from the badness of the stone, I feel it my duty to repeat again to the Committee, what I have frequently done before, the propriety of again considering whether it would not be better to use bricks generally, instead of Stone in the Works which are yet to do.

A little later John Thomas, superintendent of works, informed the committee: 'We still keep a sett of Masons repairing the Works which have been torn to pieces by the frost on the Western District.' But the Kennet and Avon company did not turn to the use of bricks despite being advised to do so by the two men most concerned with the construction arrangements. The reason, in Mr Clew's opinion, lies in the fact that much trade was expected to arise in transporting stone from Bath and Bristol to London and that the company was unwilling to offend quarry owners, who might be customers, with scenes of brickmaking. But in due course brick came to be increasingly favoured for canal works. Even quite a short tunnel needed them by the million.

In some places the lack of a proper geological survey led to the annoying discovery of unsuspected beds of rock, which could only be moved by blasting with gunpowder. Where this happened—and no brickearth was to hand—the best move was often to leave the section in question until there was a waterway on either side along which to float bricks made elsewhere.

Various enterprising commercial brickmakers, having made trial borings, took up calculated positions just ahead of an approaching waterway where the construction would need bricks; thereafter the canal, which already fed the navigators with equipment and building materials, would give the brickmakers a good means of distributing their weighty products to house-building customers. Later, several brickmakers showed the same kind of initiative in establishing themselves at places suitable for serving the railways.

While ordinary porous bricks made from canal diggings were quite satisfactory for wharves, warehouses and convenience bridges—and also for lining the ceilings of tunnels—the bricks whose job was to hold water had to be heavier and denser. The still famous blue bricks of Staffordshire were in great demand for canal work.

Because of their hardness these engineering bricks were employed for the inclined planes which occasionally took the place of locks and lifts to link sections of waterway in hilly country. In 1940 a Shropshire farmer found it impossible to dig out post holes for a cowshed on a particular site and proceeded to uncover what proved to be an inclined plane for a long abandoned canal. It was of blue bricks made at West Bromwich; its useful life, the farmer learned, had been between 1788 and 1816.

Inclined planes (or rollers) were always a last resort in canal construction because of the labour and inconvenience of surmounting them: England's first—completed on the Ketley

68 *below*
Sixteenth century.
Nathaniel Lloyd wrote that 'this beautiful house is unique and may be regarded as presenting the summit of terracotta work in England' – Sutton Place, Surrey, 1517. Terracotta consists of refined brickearth, well fired but unglazed. It was used for ornament, especially between 1520 and 1540 and in the nineteenth century

69
Seventeenth century. One of the first buildings for which rubbers were employed. Soft bricks were sawn to the shape needed by the design and then so rubbed by another brick that they formed the finest possible joints—the Dutch House, Kew Palace, London, 1631

Canal, Shropshire, in 1794—had a rise of no less than 73 feet. However, an inclined plane linking a quarry at Conkwell in Wiltshire with the Kennet and Avon Canal ingeniously took advantage of the force of gravity and was so arranged that, as a full container of stone descended, an empty one was drawn up by means of a connecting rope.

Another important building material for canals in pre-cement days was lime for mortar; and this too might be a by-product of digging. Lime excavated near Worsley (Sutton lime) had a remarkable ability, Mr Malet has pointed out in *The Canal Duke*, to set quickly in wet conditions.

In 1760 the Duke of Bridgewater's engineer, John Gilbert, noticed, while making a cut, a material known as lime marl lying about a foot below the surface. The discovery was to save his employer many thousands of pounds. Up till then lime needed for this particular canal's brickwork had been brought almost thirty miles from Buxton in Derbyshire. Gilbert was able to have the lime marl so treated by burning that it was of service both for mortar and for mixing with brickearth to make strong yellow bricks of London-stock type. He entered in his ledger:

> 29 Nov. 1760
> Paid Jos. Adking for making 275,845 bricks, £111-13-6.
> 13th Dec. 1760
> pd. Bradbury & Co for making 164,576 Bricks and 36,000 Lyme Brick, Burning as per Act and rect. except £5 kept in hand to see how the last brick proves, £104-12-3.

A cheerful result of finding the lime marl was that Gilbert was paid two years' arrears of salary and that the navigators received arrears of pay amounting to a total of £400.

Just as bricks were made at the canalside, so material that came to hand by the trackside was often converted into bricks for the railways that gradually removed the canals' trade. In

the earlier decades of railway building immense numbers of bricks were needed: 300,000 for an ordinary road bridge, for example, and 14 million per mile for a tunnel. Fortunately it was possible to make strong bricks with easily portable equipment. Charles T. Davis, clearly pleased with a design for a folding-up brick barrow, shows three diagrams of it in his *Treatise on the Manufacture of Brick* and devotes two pages to explaining its operation. A barrow of the kind was already often employed, he says, where the construction of railroad works made it necessary for brickyard plants to be often moved.

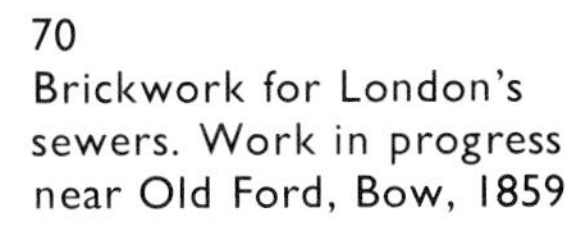

70
Brickwork for London's sewers. Work in progress near Old Ford, Bow, 1859

The Welwyn Viaduct, which was built for the Great Northern Railway Company in 1850, took 13 million red bricks which were made from the local clay. The contractor, Thomas Brassey, assembled for the brickmaking scores of labourers, some of them French, who lived rough on the site. His handsome viaduct, since refaced in yellow stock bricks, is 520 yards long and 98 feet above the surface of the River Mimram.

In the years 1849 and 1850 it was a popular entertainment for Londoners, especially in the evenings, to watch the bricks being made for the Copenhagen Tunnel, just north of King's Cross. As the tunnelling navvies excavated earth and clay, so the contractors, Pearce and Smith, made bricks of it at the tunnel mouth. Edward Dobson, author of the treatise on bricks published in 1850, watched how they did it.

> The clay is neither weathered nor tempered, but as soon as dug, is wheeled up an incline to the grinding mill. It is mixed with a certain proportion of sifted ashes, and, passing between rollers, falls into a shed, whence it is, without further preparation, wheeled to the moulders. The moulds are of wood and the process employed is that known as slop-moulding. . . . The bricks thus made are of an irregular reddish brown colour, and of fair average quality.

A hundred and twenty years later nearly all these bricks of

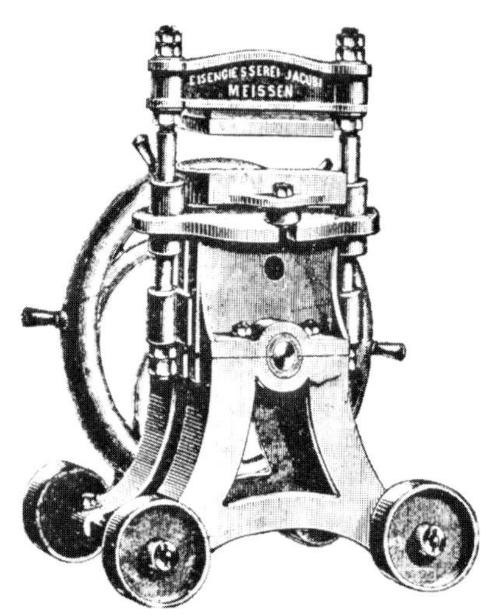

71
A German re-press machine to give bricks extra hardness. Making bricks on site for engineering works called for portable machinery

tunnel-earth, many of them uneven because of hurried handling, are still doing their job. A British Railways engineer who is familiar with the Copenhagen Tunnel says they are very good bricks indeed. Replacements have been made here and there to put right the damage done by the sulphurous smoke of steam trains, but always it is the mortar and not the brick which is found to have decayed. Today concrete sections are employed for tunnel work, but as recently as 1913 the Pondsbourne Tunnel at Cuffley in Hertfordshire was lined with bricks hand-moulded on the spot. Most of the large diameter sewers of London and other cities are still lined with brick.

There were plenty of occasions, needless to say, when the earth arising was less than suitable. Securing the raw material for the number of bricks needed for a particular work normally depended on the initiative of railway engineers who, knowing they would shortly need bricks, were able to recognise a good clay when they saw it. There are still some brickworks in Britain which owe their existence to an enterprising railway engineer.

The good appearance of brick for bridges over railways and rivers is evident to all, inviting favourable comment when compared with certain concrete bridges. Putting down bricks as the surface for roads and streets, as distinct from pavements, was rarely done in Britain in the times before tar macadam—when such paving would have been useful. Why is not certain. A brick-on-edge service road at Redlands' Southwater Brickworks in Sussex has needed no repairs since it was laid in the early 1930s.

Bricks for paving roads were popular for many years in parts of America despite the distances they had to run: as early as 1719, according to Charles T. Davis, they had been in great demand in Philadelphia for sidewalks. In 1914 the *Brick and Pottery Trades Journal* reported:

> The success of the brick road in America has been largely due to its permanence under the new conditions of power traffic. The hardness of brick enables it to withstand the friction and tear of the rapidly running automobile. . . . Such roads have won their way to the appreciation of auto owners and farmers alike, being equally passable in any season and so drained that every rain washes them clean.

The quality of brick roads so appealed to Governor Glynn of New York State, the journal went on to say, that he had devoted an entire message to the legislature advocating brick for all main highways in the State. Cuyahoga County, Ohio, had already nearly four hundred miles of brick roads, apparently—built at about $1,000 a mile for each foot in width. All have now disappeared beneath tar macadam or concrete. In Holland on the other hand roads of brick are still often encountered. These are traditional as well as efficient. Hyde Clarke, writing about the Dutch brick industry in 1850 for Dobson's *Treatise on Bricks and Tiles*, reported that the Dutch used bricks—especially the hard clinkers—for foot pavements, canal towing paths, streets and high roads; the 'slime' for clinkers came from Haarlem Meer and was collected by men in boats equipped with long poles that had a cutting circle of iron at the end.

72
The Railway Viaduct at Digswell, Hertfordshire, 1850. It took 13 million bricks which were made on the spot from local clay

In recent years brick has been increasingly in demand in Britain and on the Continent for paved walks in cities and traffic-free shopping areas. The authorities choose it, as it is chosen for many interior walls and stairways in commercial buildings, mainly for its aesthetic appeal.

Transport 13

The labour of carrying bricks over bad roads by horse and cart—not more than 800, or two tons, at a time—largely accounts for the multiplicity of tiny and temporary brickyards in the days before railways revolutionised the transport of freight. The difficulties made it good sense for anyone putting up a large building to have his bricks made within feet of the work if reasonable clay was to hand: when the Earl of Nottingham decided to build in brick at Burley-on-the-Hill, Rutland, he sent to Middlesex for men skilled in brickmaking to work for him on the spot.

The only comfortable way of transporting bricks a long distance was to float them. Some of the mediaeval bricks to be seen in the eastern counties of England came slowly across the North Sea from Flanders, and it appears that a few may have

73
Loading horse-drawn carts in the early nineteenth century

been taken on along the south coast and up the River Exe—to judge by the existence at Topsham in Devon of characteristically small Dutch bricks. It has been supposed that Dutch bricks reached England as ballast, but Miss Wight in expressing doubts about this in *Brick Building in England* is not alone. 'Trade, especially with North Europe,' she writes, 'was too regular and intensive for there to be wasted space on the return journey to be filled with bricks as ballast. The bricks would anyway have to be purchased. . . . The idea of odd batches of bricks being used for ballast and then being used in England for building seems implausible, whereas speculative or ordered cargoes of bricks do not.'

Transport difficulties partly account, too, for regional variations in the houses of pre-nineteenth-century England which observant travellers enjoy noting today. Local building materials differ in characteristics from county to county—and in the old days they stayed where they were with distinctive results. The warm mellow bricks of the Sussex Weald are not to be confused, as Mr Clifton-Taylor has pointed out, with the rich purple bricks of East Dorset, nor the pale tones of the Fens with the brilliant reds of the Midlands. The greys of Bedfordshire were as distinct from the yellows and brimstones of the Thames Estuary as were, later on, the bright blue bricks of Stourbridge from the white-ish bricks of the Severn Valley, which yielded white clays with no iron and a lot of silica. Regional differences in buildings were only marginally blurred by sea transport. The coming of canals, however, had a noticeable effect.

Brindley's great achievement, towards the end of the eighteenth century, of bringing by artificial waterways the Duke of Bridgewater's coal from mines at Worsley into Manchester began an epoch of canal building throughout the country. Over 3,000 miles of canal were made, and for the first time in Britain's history it became possible to transport heavy goods economically.

The bricks of one region could now be taken into regions at a distance and, of course, into the heart of the stone-producing counties. Industrial undertakings springing up in various parts of Britain needed huge amounts of material for their buildings, and bricks, always cheap to make, could now be cheap to carry.

The canal system, as it grew in the early nineteenth century, itself encouraged brickmaking and clusters of small brickfields appeared at strategic points. In *The Last Chronicle of Barset*, 1866, Trollope described a West Country stretch of canal giving access 'in some intricate way' both to London and to Bristol. On its banks there had sprung up

> a colony of brickmakers, the nature of the earth in those parts combining with the canal to make brickmaking a suitable trade. The workmen there assembled were not for the most part native-born . . . they had come thither from unknown regions as labourers of that class do come when they are needed . . . they were all in appearance and manners nearer akin to the race of navvies than to ordinary rural labourers. . . . The farmers hated them, and consequently they hated the farmers. They had a beer shop, and a grocer's shop and a huxter's shop for their own accommodation. . . .

The important firm of Eastwoods, which began to build up a chain of stock-brick yards in north Kent in the early 1800s, owed much to having access to canals and rivers via the estuary. This firm made cement as well as the admired buff stocks fired in open clamps, and had within its grasp a market which included expanding London. If Eastwoods could not make bricks quite as cheaply as the fletton makers around Peterborough, at least they had the edge on their rivals for many years in being able to transport their products at small cost.

Details of the early history of Eastwoods Ltd are given by Mr Willmott in *Bricks and Brickies*. In 1855, he relates, they ceased to rely on chartered water transport and began to build sailing barges of their own. Three were built during 1862 alone: *Arthur and Eliza*, *Rapid* and *Neptune*. Eventually Eastwoods had a fleet of seventy sailing barges for which late in the nineteenth century annual races were staged; a barge called *Surrey* was prominent in the film *Red Sails in the Sunset* starring Jessie Matthews.

The firm had its own house flag, or bob, which was always proudly fixed to the truck of the topmast. The bob was red, white and blue and, set against the rest of the colours chosen for the barges, it capped a pretty spectacle: hulls were black, rails light blue, and bow and quarter boards brown; there was a green transom. The fleet was not finally dismantled or sold until after the Second World War. Sometimes the barges were referred to as brickies, which is also a friendly term in north Kent for those who make bricks; elsewhere only for those who lay them.

The return trip from London with fuel in the form of ashes, coke and rubbish was always unpopular because of the chance of spontaneous combustion bringing carbon dioxide fumes, and crews watched a lamp, knowing that if its flame grew low it was time to go up on deck. The men became expert at navigating canals. Barges working the Regent's Canal were not permitted a length of more than 78 feet (inclusive of rudder) and were not to be loaded to a draught of more than 4 feet 6 inches. The restrictions nevertheless allowed a barge to pass through thirteen locks up to the Grand Junction. There were two tunnels on this run of eight and a half miles; they were known as legging tunnels because to propel the craft through them the crew lay on their backs and pushed against the roof with their feet.

Eastwoods had a wharf up the Surrey Canal where men seemed always busy handling bricks. Barges working this canal were not allowed a beam of more than 17 feet 6 inches. In the early years barges could enter the Surrey Commercial Docks only by the middle, or Lavender, entrance, a circumstance which gave

74 *left*
Loading barges—a photograph of 1905

75 *right*
The successor to the horse and cart, 1900

crews anxious minutes because of the road bridge that ran over the cutting: to pass under it took skilful handling, especially when a strong tide was flowing.

The smaller brick barges sometimes negotiated the Gravesend and Higham Canal, which was for a long time owned by a railway company. After entering the lock gates at Gravesend at high water, they were towed by a horse at a charge of 6s. each way. A beam of not more than 15 feet was permitted on this canal and loading was not to exceed a depth of more than 4 feet.

Horse-drawn carts were for obvious reasons still needed in abundance—to carry bricks from a vessel's unloading point to the building sites; and they were equally essential at these points when the railways came. Wherever possible rail transport was universally made use of from, say, 1900 to 1930. Nearly all the big brickworks built around the turn of the century were

MANN'S PATENT STEAM CART & WAGON CO., LTD.,

Canning Works, Dewsbury Road, LEEDS.

Telegraphic Address: "CANNING, LEEDS." A B C Code used.
Telephone: No. 1335[1]

Also made with Flat Lurry Bodies, 12ft. long.

Specially adapted for Brick, Stone, Coal Leading, &c.

Patent Steam Tipping Cart with Steel Body.

Constructed to come under Motor Cars Act and carry up to 5 tons. No License required and only one man in attendance. Burns ordinary gas coke. Does not damage roads. As handy as a horse and cart.

sited near a railway, and had private sidings which were used not only for the dispatch of bricks but also for the receipt of coal. Brick barrows were wheeled directly from the kiln into the railway truck.

However, a change in the transport of heavy goods was foreshadowed at the end of the nineteenth century by the development of steam traction on the roads. Use of the new steamers for a complete journey meant that there need be no intermediate off-loading and reloading. When Eastwoods supplemented their fleet of barges with a steam traction engine they found, with mixed feelings, that a load could be delivered to any part of Kent in one day—a 'load' consisting of 1,000 bricks weighing $7\frac{1}{2}$ tons distributed in three trucks.

The journeys called for some resourcefulness. For example, in taking bricks to Maidstone, the trucks had to be hauled up

Detling Hill one at a time. A man rode on each towed truck, linked to the driver by a piece of cord tied to his fingers; tugging at this was usually a signal to move over to let another vehicle overtake. The driver needed the signal because he could hear little above the clanking of wheels and hissing of steam.

The official speed limit for steam traction engines was four miles an hour, but a former driver for a brick firm (he thoroughly enjoyed his work) has told me that they commonly reached twenty miles an hour. Improper speed especially accounts for the fact that steam traction engines running on the unmetalled roads of the day often got into trouble for breaking up the surface. By 1900 the damage being done to roads by these engines was a nuisance which had become a talking point.

A correspondent to the *British Clay Worker* suggested that the way to avoid damage was to do without the cross bars fitted for gripping purposes to the road wheels of the engine and, further, to carry the load instead of drawing it—with a 10 ton load of bricks immediately behind the engine there would be good adhesion to the surface. And if only wider wheels could be employed, the writer declared, there would be a rolling effect and the wheels would actually improve road surfaces by their passage. The bars remained. Wide-wheeled steam traction engines in the form of huge lorries sometimes did serve to roll a road.

Prosecutions were numerous. The judgments were worrying because of the need to base them on opinion rather than facts. William Atkins, a brickmaker of Mountsorrel, was obliged in February 1900 at Leicester County Court to pay the county council £105 towards repairs to the main road between Leicester and Loughborough. He was found guilty of causing extraordinary traffic on this road during the previous year, with the result that deep ruts had appeared in it.

Atkins told the judge that until two years before he had sent his bricks from Mountsorrel to Leicester in carts, each pulled

76
Transport by Fletliner train in 1973—Mr Jeremy Rowe of London Brick posing on a consignment of flettons

by a single horse; the loads had been about 2 tons including the weight of the cart. Although he had not increased the total number of bricks dispatched, he had changed his mode of conveyance by buying a steam traction engine. This engine weighed 11 tons, and by means of it he had been drawing his bricks along the highway in big trucks, each of which when loaded weighed about 10 tons. The judge considered that this method of haulage put excessive weight on the road within the meaning of the Highways and Locomotives Act, which made people liable to pay for damage done by 'extraordinary traffic' and 'excessive weight'.

Road steamers and a growing race of bicyclists and tricyclists at length impelled the authorities to carry out radical improvement work on the roads, which had been neglected during the years of rail travel; and within a surprisingly small number of years all main routes were sealed with tar and stones. Steam transport on the better road surfaces remained a familiar sight till the 1930s, though petrol or diesel-driven lorries were in business before the First World War. These early ones had solid rubber wheels and a chain drive and they hardly carried more bricks than a horse and cart. In 1948 Eastwoods built up a fleet of lorries with a long wheel base which could carry 7,000 bricks. On the other hand a brickyard at Debenham in Suffolk, which closed in 1936, never owned any transport. The proprietor's daughter, Mrs Nunn, who was tape-recorded by a British Brick Society member, recalled that 'the customers would come to carry their own bricks away' and that 'they came from all over the place—Earls Soham, Wetheringset and all round that way.'

The convenience of lorries that can swiftly convey bricks from door to door, so to speak, greatly reduced the part of railways in their transport. There is now, however, a sign of change since the introduction of a container service on the railways which eases the labour of transferring loads.

One of the results of better transport in recent years has been

77 *right*
Eighteenth century.
A large proportion of blue headers emerged from the wood-fired kilns of Sussex. They are seen here to advantage in cottages and a house at Uckfield, Sussex

78 *above left*
Twentieth century.
A mud-brick, or adobe, hut in Mexico

below
Building with mud bricks at Hadhramatet, Aden, in 1971

that the inexpensive bricks of, for example, London Brick Company, are to be seen on building sites in every corner of the British Isles. But these mass-produced bricks of the fletton areas need no longer look wrong in certain surroundings, for they are available with a mellow finish calculated to let them fit in anywhere, even in stone districts. There has emerged from the laboratories of the brick firms a whole range of pleasing colours and textures that can be applied to common bricks. London Brick's Cotswold and Dapple light varieties, for example, are calculated to match traditional building materials and look well in new housing in the Cotswolds and the North Riding of Yorkshire. Redland Brick's cheap wirecuts known as Crowhurst pastone rustic and Witley pastone rustic, which are scored to give the creases of handmade bricks, look at home among ancient buildings and are no affront to a predominantly green background.

Still, however, modern transport allows some places to receive bricks and other units which clash with their surroundings. Still there is a lack of thought in the choosing of building materials. Bright red tiles have appeared on the roofs of new stone-built houses in Bath.

Mr Clifton-Taylor remarks about roofs: 'It is a strangely twisted mentality which can indulge in the erection of a shed roofed with pink asbestos tiles on the edge of a slate quarry. Yet this is what can be seen today at a place in Cumberland.' He recommends that white asbestos sheeting on farm buildings should be coated with a bituminous paint of dark bronze green, for 'it is a basic aesthetic truth that a roof which is darker in tone than the supporting walls imparts to a building a feeling of repose, so that it seems to sit more firmly on its site.' This seems to explain why houses capped with thatch, which is pale, look at their best when the walls are white, softly colour-washed or of pale light-coloured stone. The makers and choosers of bricks have something to learn from old non-brick buildings.

The fletton brick 14

A revolution in brickmaking occurred in the 1880s when bricks began to be made much more cheaply with a tough shale-clay dug up at the village of Fletton near Peterborough. Brickmakers found that it burned almost of its own accord because of a 10 per cent fuel-oil content. This previously undiscovered material, known geologically as the lower Oxford clay, stretches in a deep band some feet below the surface from Yorkshire to Dorset.

For efficient working, coal dust was sprinkled among the bricks, but with care the cost of fuel, even in primitive kilns, was not more than a third of that incurred elsewhere. Lower Oxford clay, coming from the ground with a modest water content of about 18 per cent, had the equally important property that it could be formed into bricks that were ready to be taken straight to the kilns, without any preliminary drying.

Fletton bricks were made from the start by a process known as semi-dry pressed. To break a fletton, and expose the wholemeal-bread texture of bumpy granules, is to see one characteristic effect of a process in which, instead of working clay made plastic with water (this water must later, of course, be removed) the brickmakers crush up the semi-dry clay just as it comes and force the powder to cohere by applying great pressure. Today some 3,500 million British bricks are made each year in this way.

The brickmaking qualities of lower Oxford clay was a discovery, like finding gas under the North Sea, which has had far-reaching effects. Without the clay, without a continuing source of bricks that were cheap, it is likely that Britain would long ago have resorted to the Continental practice of building with hollow clay-blocks which need rendering to keep out the weather and to disguise them: these blocks are laid in a rough and ready way for, as German builders remark, 'Der Putz deckt alles'—the plaster covers everything. Britain had been for centuries a land of good solid brickwork: lower Oxford clay enabled the conventional brick to remain the builder's friend.

The story of fletton bricks begins in 1877 when the Fletton Lodge Estate was sold. The auctioneer announced laconically that there was brickearth on the site, thereby drawing the attention of local brick firms, and within a short time there were six small brickfields on the land turning out bricks by the wirecut process. But the advertised brickearth was a superficial deposit of ordinary yellow clay.

The great discovery was made in 1881. A firm called Hempsted Brothers decided to try using the material which lay below the yellow clay and which was so hard that it brought up short the teams of diggers. Trials were promising and Hempsteds ordered heavy grinding and pressing machines from Lancashire, where the semi-dry pressing of another kind of shale was already established. The fletton brick had been born.

One of the first problems was how to get the bricks fired more efficiently; for large outputs were causing an acute shortage of kiln space. The chimney-less intermittent kiln, suitable for small undertakings, was inadequate for burning in bulk the new type of brick. Also, because water was extracted in the kiln, firing bricks of lower Oxford clay caused a great deal of smoke.

The situation was brought to a head by a strong but friendly complaint about fumes from a wine merchant, James Bristow, who lived beside Hempsted's works at Fletton Manor.* Having taken the best advice, Hempsteds invested in a modified Hoffmann continuous kiln which both dealt with the local nuisance of fumes and allowed a much greater output of burnt bricks.

* The late L.P. Hartley, the novelist, lived in this house in his youth. He was the son of a brickmaker.

79
Ten million bricks of lower Oxford clay were supplied for the building of Westminster Cathedral in 1898. The order marked a turning point in the fortunes of the fletton industry

Mr Bristow, their mollified neighbour, later left the wine trade to make fletton bricks himself; his son, grandson and great-grandson also became brickmakers.

In the Hoffmann kiln—as explained in chapter 11—the combustion gases pass round from one chamber to the next and are to an extent cooled by the time they are drawn up the chimney. Hempsteds further dissipated their smoke by erecting a chimney much higher than was then customary. Ever since, chimneys between 100 and 400 feet in height have been a feature of the fletton industry, and clusters of them today dominate the rather flat parts of Bedfordshire and Cambridgeshire where the main plants of London Brick Company are situated.

Too many of the new fletton brickmakers lacked the finance to operate properly; yet all had to compete against one another. Because of undercutting and price wars, the course of trading long remained anxious and studded with bankruptcies; and it is doubtful if any participant in the fletton industry grew rich in the first twenty years or so. An average ex-works price of 22s. per 1,000 in 1896 dropped by degrees to the suicidal figure of 8s. 6d. per 1,000 in 1908.

As well as worrying about what their neighbours were doing, the fletton makers had to face severe competition from brickmakers who produced 'soft mud' stock bricks in Kent and Essex. These contrived to bring their prices right down, and in the process reduced wages: even the all-important moulders commonly received only 2s. 10d. per 1,000 in Edwardian days—nearly 2s. less than had been usual in the 1880s. But labour was willing and the occasional local strike generally ended with a return to work without the extra penny or halfpenny being conceded. Fletton brickmakers also had to overcome some initial resistance to the use of their bricks. Their competitors fearful of the growth in fletton sales suggested that the fletton brick was not strong enough for normal building work and would deteriorate when used on foundations. Only gradually as experience of flettons grew, were such prejudices overcome.

Firms that had to shut could recoup little from selling their land, which fetched perhaps £20 an acre: there might be lower Oxford clay ten or more feet below the surface, but it was having the means to exploit it that counted. When a group called the New Peterborough Brick Company was formed in 1897, and offered shares to the public, the *Pall Mall Gazette* commented: 'We cannot imagine anyone subscribing . . . the issue should obviously be left alone.' This group nevertheless had the necessary machinery and three Hoffmann kilns.

Yet all the time the fletton makers were doggedly producing more and more bricks. The total of about 50 million made, and sold, in 1890 had increased ten times by the end of the century, by which time there were already sixty tall chimneys in the lower Oxford clay area. In 1898 the firm of Shillitoe fulfilled an order of no less than 10 million flettons for parts of Westminster Cathedral, and two years later the *British Clayworker* announced: 'Fletton is a force that must be reckoned with. Its amazing strides are one of the commercial wonders of the century.' Between 1903 and 1904 some 25 million flettons were used in the new War Office building in Whitehall. They were sold, though, for the low price of 27s. per 1,000 delivered to the site. The *Building News* commented: 'No one could make a profit on that figure.'

In 1901 a mechanical clay digger was tried out in the fletton fields for the first time—it had done service in the Manchester Ship Canal—and showed the extent to which the operation of the pits might be speeded by mechanisation. It was becoming clear that if anyone was to produce flettons economically, more and better machinery was required; and to have this meant belonging to big, properly financed groups. But scores of small fletton firms continued to struggle along.

The rub came in the building recession of the 1920s when many of the small makers found that they could not compete with the larger firms in the industry. London Brick took the lead in organising a series of acquisitions and mergers designed to

80
A shale planer built in 1926 winning lower Oxford clay for the manufacture of flettons

rationalise production and introduce economies of scale. As smaller companies were acquired, so new machinery was installed in their works and their bricks sold through a national marketing network. By the coming of the Second World War, the Company had revolutionised the fletton industry and in so doing had become the largest single brickmaker in the world. In the 1960s and early 1970s, London Brick acquired the remaining two major fletton brickmakers, Marston Valley and Eastwoods, and in 1973 finally became the sole fletton producer when it bought from the National Coal Board the two remaining fletton works at Whittlesey not previously under its control.

There is little doubt that London Brick grew to dominance not only because of its sounder financial base but because, of all the small firms that had grown up in the early days of the industry, it alone saw the potential that existed in fletton brickmaking for large scale mechanisation and the need to promote the product on a national scale. Today, London Brick is one of our major industrial companies which apart from making half the clay bricks used in Britain, operates successfully in other associated industries. Its Directors, however, still have their roots very much in the fletton brick industry.

The history of London Brick goes back to 1889 when J.C. Hill bought a brickworks at Fletton for £6,500 and called it The London Brick Company. But 1900 is a date of particular interest to the company because it saw its incorporation as London Brick Company Limited and also the incorporation of B.J. Forder and Son Limited with which, in 1923, it was to amalgamate.

Forders, in beginning their main operations with lower Oxford clay, settled themselves beside the important railway line at the Bedfordshire village of Wootton Pillinge. It was Britain's fully developed network of railway lines which was to launch the cheap fletton and make it a national product, and Forders took full advantage of their line, sending down it many train loads of bricks to the rapidly developing suburbs of London.

The Forder group, starting with capital of £280,000, was able to install the latest steam-driven machinery, of which the Forder long jib clay excavator—still working twenty-five years later—was the most spectacular. A system of steam-driven endless chains hauled clay wagons to the making sheds and permitted the disappearance of crude winches. New brick presses were alarmingly powerful and an accident common until the coming of safety devices involved damage to operators' fingers. These machines were so adjusted that they gave each green brick four pressings: the trademark 'Phorpres' coined in 1901, was till recently to be seen impressed within the weight-saving and theoretically strengthening frog (or hollow) of nearly all flettons. Grinding pans installed to crush the knots, as the workmen called the shale-clay, were so efficient that when in 1897 a boy fell unseen down a feeder chute, he disappeared, no one realising what had happened until fragments of bone were noticed in the ground clay going to the presses.

The first chairman of B.J. Forders, and the man who largely financed the new equipment, was Halley Stewart, a preacher and politician and a businessman who had made a fortune already out of cattle cake. His fellow-directors expected quick returns on the ground that their fuel bills were trifling compared with those, say, of the Birmingham brickmakers, who had to pay more for coal than for having the bricks made. But Stewart was determined that profits should be ploughed back until satisfactory reserves had been built up. That might take ten years, they protested. Even if it did, replied Stewart, there was no other way to build a successful business in bricks. In fact fifteen years passed before Forders paid any dividend, except on the preference shares.

As a rich man, Halley Stewart was in a position to wait. He enjoyed the business. Making bricks—bricks to build houses, houses to provide homes—appealed to his urge to serve the public. He had with him as fellow director his son Percy Malcolm Stewart. They worked well together, particularly in that Percy bowed to his father's financial decisions.

Forders tried from the beginning to charge an adequate price for their bricks. After the forming of the Pressed Brick Association in 1909, in which makers got together to put an end to ruinous price cutting, the average price for flettons was 13s. per 1,000. The company also paid the men better than for comparable jobs in the neighbourhood and provided eight cottages and a large hut for some of them to sleep in. Men could earn up to 28s. a week. Labourers on the farms getting about 12s. a week were attracted to the Wootton Pillinge brickworks by the extra money and, indeed, two brickworks in the area were nicknamed Klondyke and Kimberley.

By the end of 1910 it began to seem that the management, too, was at last striking gold in lower Oxford clay. Forty-eight million fletton bricks had been dispatched in a year, an immense number for those days.

The works became almost derelict during the First World War, but the great demand for bricks afterwards stimulated fast re-equipment. In 1926 no less than 118 million bricks were made, despite the fact that the day and the night shift now worked only 48 hours a week instead of 56½ before the war. Five hundred people were employed. The *Investor's Review* was to praise Forders for having 'applied itself from the outset to reducing the import of foreign bricks from countries where labour conditions compared unfavourably with British conditions'.

In 1936 the annual production at an extended version of the one works had risen to 500 million bricks; it employed 2,000 people and could claim to turn out more bricks than any other works in the world. In that same year Bedfordshire County Council agreed that the village of Wootton Pillinge should have its named changed to Stewartby as a tribute to the achievement there of Halley and Percy Stewart.

The rise of the cheap fletton in this period, and the cyclical nature of house building, led to numerous small brick works being shut down. Despite the cost of transport (this was often reduced by favourable contracts with the railway companies), the scale of production made it possible for trainloads of flettons from the Bedford, Peterborough and North Buckinghamshire districts to compete hundreds of miles away with bricks made in local yards. In 1900 there had been about 3,500 brickyards in Britain. By the start of the Second World War the number had been reduced by at least two-thirds. Today, after further waves of closures of non-fletton works in the 1940s and 1950s, the number of brick concerns is about 350.

In the second half of the 1920s those who worked under the better conditions provided by the firm of London Brick and Forders began to realise that brickmaking need not be—as it had always been before—a rough-and-ready and often degrading trade. The notion spread, and people were applying for jobs for reasons other than wages slightly higher than in farming.

Until 1926 there had been at the biggest fletton works little of the paternalism now always associated with London Brick and valued by the company's work force. People had been abruptly dismissed, as at all other brick works, according to the state of trade. There were no rooms to have meals in; wash places were primitive; first aid arrangements and lavatories barely existed. But in 1926, Percy Stewart, who had become chairman two years earlier, set about a far-seeing programme of looking after his workers and their families. He introduced a week's paid holiday and a profit sharing bonus system—both considerable innovations for the 1920s—and he recognised the trades unions. Canteens were built in which decent meals could be had.

Percy Stewart put in hand the building of sixty-six carefully-designed workers' houses near the works, each with a back boiler and a small garden. In the early 1930s more houses were built and, to go with them, community buildings and sports

81
London Brick uses common flettons for its tall kiln-chimneys. These men are working 116 feet from the ground

grounds. Swimming pools were installed. There are today over 1,000 people in the model village of Stewartby, living in 358 houses and bungalows. There are two schools (they come under the control of the county council) two general shops and an interdenominational church. On being made a baronet in 1937 Percy Stewart used his second Christian name and became Sir Malcolm Stewart of Stewartby. His son, Sir Ronald Stewart, is a well-liked chairman and his employees so rarely leave that gold watches for long service are bought in bulk. Over 1,600 have now been presented for twenty-five years' service or more.

Many of the houses at Stewartby are built (unfortunately in stretcher bond) with the company's Rustic bricks, a well-known type which is scored with wavy lines and has a uniform reddish colour. Rustics, the first flettons to be offered for facing purposes, were invented by the company in 1922. Until the 1930s all other flettons were common bricks intended not for visible parts of buildings but for foundations, backing-up work and for walls that were to be rendered or colour-washed.

The fact that in practice these common flettons, pallid and heavily barred from strategic (or economical) placing in the kilns, have been used not only for the backs of cinemas but for the exterior of whole buildings is visually a pity. Mr Alec Clifton-Taylor refers regretfully to 'acres of pasty-faced Flettons' in the 1963 edition of his *Pattern of English Building*. In an edition nine years later, however, he absolves the makers from blame—customers had misused these bricks—and draws attention to the mellow, faced flettons on the market.

LBC products are indeed transformed by a facing treatment applied to the surface of the brick. This may consist of mechanically roughening the surface or sandblasting it before firing and thereby hiding the wide bars known as kiss marks. In practice, several minerals besides sand—powdered slag, powdered brick—are sprayed on for particular effects. It was once found during experiment that ground-up road sweepings gave a pleasant finish, and a laboratory assistant's trial with a handful of steel slag, noticed during a lunch-hour stroll, brought into being the popular grey brick with speckles called Dapple Light.

The range of fletton facing bricks has been developed to suit the particular local traditions in British building. The Dapple Light and Cotswold facings harmonise in areas of predominantly stone building, the Tudor, Heather and Sandfaced bricks with the red brick tradition of Kent and Sussex. Even the London Stock brick has been copied in the Golden Buff and Milton Buff bricks which look well among the older brick buildings of London.

The method of making bricks is similar at all London Brick's plants. The clay reaches the making sheds by conveyor belt—at Stewartby 600 tons of it an hour—and is put through crushers and grinders. On reaching banks of electrically heated piano-wire screens it has fragments above a certain size rejected for regrinding, while the fine granules are fed into press hoppers. In each press a charger working backwards and forwards fills a mould box with a measured charge. Two pistons, one above and one below, compress the clay powder, pause to let out air, and compress it again. A brick has been formed. It is then pushed forward into another box and pressed twice more to give the advertised total of four pressings. Bricks which are to have a decorative facing leave the presses on rollers and are transferred to a single-row belt to be moistened to make them tacky and then sprayed with a sandblaster.

Since 1948 setting and drawing bricks in the kiln has been done by means of fork lift trucks which are fitted with a clever device, patented by London Brick, which does away with the need for wooden pallets to support the stacks of bricks. This consists of pneumatic rubber tubes fitted between the picking-up prongs which, on being blown up, grip the bottom course of bricks where gaps have been left, thus enabling the rest of the load to be carried across the prongs. To put the load down it is only necessary to deflate the tubes so that the fork can be withdrawn.

82
A fork lift truck in action. Note how the arrangement of the bottom row of bricks enables the tines (or prongs) to pick up the load from the ground. Incidentally, the way these flettons are stacked explains the bars on them when they come from the kiln—some parts are in direct contact with heat

Although the green bricks are rigid enough to be taken straight to the kiln as soon as made, they must not be submitted at once to intense heat. The fork lift trucks—usually driven and operated with verve—set them down in those chambers of the Hoffmann kiln from which fired bricks have been taken and which are merely warm. In these chambers the bricks are gradually dried by hot air drawn from the chambers where fired bricks are being cooled.

When all the water has been extracted from the clay and the bricks are dry, the fire is introduced to the bricks which, with their carbon content, literally ignite. At 1,000°C a further rise in temperature is halted by allowing cold air to pour into the hot chamber for several hours. Once the temperature begins to drop, the easing process is stopped and the bricks are kept at just over 900°C for some thirty-five hours to remove residual carbon. To maintain this temperature small handfuls of coal dust are fed in through bung-holes in the roof of the kiln. During their subsequent cooling process, the bricks act as a preheating unit for the combustion air flowing through to the main fire and also provide hot air for the drying chambers. The total time in the kiln is about fourteen days.

83
One of two London Brick works which at King's Dyke are built unobtrusively at the bottom of a worked-out clay pit: only the chimneys are visible from the surrounding countryside

London Brick Company lorries, courteously driven, are among the least troublesome of the large vehicles on the roads of Britain. But they are very plentiful. The railways have been used less and less, even for long journeys, because of the labour of loading and unloading: goods can only be moved by rail between towns and invariably have to start and finish their journey on a lorry. A recent encouraging development has, however, halted this trend. Because of a new way of packing their bricks in containers, London Brick can now utilise the Railways' Freightliner service and transfer their bricks mechanically from one form of transport to another.

London Brick has in fact become the first manufacturing company to operate a system of sending its own containers into the Freightliner network. New sidings at Stewartby equipped with a travelling gantry crane were completed in 1973, and on 18 June of that year the first trainload, comprising 320,000 bricks in forty-five containers, set off for Manchester and Liverpool. The innovation could be as important as the fork lift truck.

The name of the new system, London Brick Fletliner, or Fletliner, was the result of a competition organised through the Company's house journal, the *LBC Review*. Within months the name was becoming familiar in the building industry. It provides a means of distributing more than three million bricks per week to London and the north-west of England.

The extraction of the Lower Oxford Clay for fletton bricks has

left the large and unsightly worked-out pits which can be seen by anyone travelling north on the railways running through Peterborough or Bedford. Because Oxford Clay was formed in the Jurassic period some 150 million years ago, the pits provide a rich and unusual harvest in palaeontological remains. Most of the skeletons of prehistoric animals exhibited at the Natural History Museum in London, were originally discovered and painstakingly recovered from the fletton brick pits. The Company's navvy drivers are no longer surprised to see fossilised remains when digging the clay and some of them have become keen amateur palaeontologists. John Horrell, the Company's Geologist has even had the distinction of having newly discovered fossils named after him.

One of the most remarkable discoveries was made in August 1973 when at a works near Stewartby the remains of an almost complete pliosaur—a vast prehistoric reptile—was unearthed. A workman noticed some large bones sticking out of the clay and told the works manager. The works manager told the research laboratory and the research laboratory rang up the British Museum. The pliosaur, identified officially as such, was presented to the Museum for its permanent collection. Some of the bones had been destroyed during excavation but everything available has been put together. The skeleton measures 21 feet across the hind paddles, and its total length seems to have been about 35 feet. Having found the largest pliosaur in the world, London Brick display a reconstruction of its skeleton in the entrance hall of their research laboratories at Stewartby.

The pits not only provide interest but a source of potential wealth to the fletton brickmaker. Over the years London Brick has become a large landowner growing crops and breeding cattle on land destined for future clay reserves. The pits themselves present a challenge: not only is it desirable to screen the actual workings by trees and landscaping, but the land must be restored to productive use. The Company has set up a new subsidiary for this purpose—to reclaim land through the operation of a waste disposal service and to redevelop other water filled pits for leisure. Pits in the Whittlesey area of Peterborough have been used in a different way. Two of the newest works have been built on the base of these, some eighty feet below the ground. This both allows the works to be built alongside the source of its raw material and screens them from normal view.

Perhaps the most ambitious reclamation project of all was publicly demonstrated at a ceremony at Fletton on 16 October 1973, when the Central Electricity Generating Board ceremoniously handed back to London Brick 112 acres of cratered land made level by dumping there four million tons of ash from the coal-fired power stations of the Trent valley. This represented ten years' disposal of surplus ash.

The ceremony marked the end of the first stage of one of the biggest schemes for reclaiming land ever carried out in Britain. An entirely practical project, it uses the waste products of one industry to make good the effects of another. By the end of the century more than 1,000 acres of old clay pits should be made level with the aid of 30 million tons of ash.

The ash, looking like powdered cement, is brought to an unloading station in sealed rail wagons that continuously circulate between the Trent power stations and the fletton brickfields. There it is mixed with water to form a slurry and pumped into the pit through 12-inch pipes. As the ash settles, the water is taken away through a circulating system. After the pits have been filled with ash and consolidated the surface is covered with about 6 inches of top soil conveniently drawn from the washing of sugar beet at a local factory and the land returned to agriculture.

It is appropriate and a matter of quiet satisfaction to the fletton brickmakers that sheep are now grazing on meadow land at Fletton from which nearly a hundred years previously their forefathers first dug their clay for brickmaking.

A brickyard on a private estate

15

Several landed proprietors had their own brickyard until well into the twentieth century. It was considered as appropriate as the estate laundry and dairy, only rather more interesting to the children. Queen Victoria's children seem to have spent a lot of time in a brickyard within the Osborne estate on the Isle of Wight. They made bricks themselves there, under close supervision, and in particular the red and blue bricks for a miniature fortress and barracks (builder: Prince Arthur, ten) which is today on show to the public behind the Swiss Cottage.

The lives of those who worked in most of the estate brickyards were secure and pleasant compared with those of brickmakers outside. Their employer or his agent may have sold bricks commercially from time to time, but the main purpose of the brickyard was to supply the materials for farm buildings at home and for repairs and additions to them. The men may have worked very hard for much of the brickmaking season, and they may have been paid according to the number of bricks made, but they did not have to get their wives and children to act as assistants or worry about making ends meet during the winter. For the estate brickmakers, housed in their employer's cottages, there was at all times of the year some kind of cleaning-up or maintenance work to be done.

An interesting example of a small estate brickyard was at Ashburnham in Sussex. Although now closed, it was in operation until 1968—and brickearth there had been worked continuously since the fourteenth century. The yard's hand-moulded bricks were fired with wood, which meant there was plenty of winter work gathering the enormous quantity of fuel needed for the summer and looking after the woodlands. Two of the men who

84
Brushwood stacked ready for firing the estate kiln at Ashburnham, Sussex, in 1968

worked in this yard up to when it closed have told me glowingly of the experience; they were proud of the bricks and tiles they made and still got pleasure in looking at these in the walls and roofs of buildings all over the neighbourhood. They especially admired the pink and grey brickwork of Forge Lodge, Ashburnham.

According to an account by Mr Kim Leslie in *Sussex Industrial History* the Ashburnham brickyard was moved every century or so so that a new site for digging brickearth could be taken. The final yard was prepared in 1840, and the Steward's account book for that year, and for the year 1841, records as follows the erection of new buildings and the stocking-up with fuel:

1840

	PAID	£	s.	d.
October 12	Jas. Colman labor to new brick kiln	2	14	0
	Jno. Winchester labour in new yard	3	16	1½
Novr. 14	Jas. Barden do	2	12	6
	Jno. Billings building Clayhouse	2	2	6
	Brick Duty	7	13	1½
	Jno. Sinden for Kiln Faggott	2	5	0
	Richd. How carrying matrl. to new brickyd.	5	8	6
Decr. 7	Brick Duty	3	13	6
1841				
January 7	Saml. Cornford Arch Bricks	3	12	0
11	Thos. Hobday Carrying Clay sand &c	9	4	0
14	Thos. Croft for Kiln Faggott	1	2	6
	John Shaw for Bricks	1	14	6
15	Jno. Isted Carrying Kiln Faggotts	12	10	11
Febry. 8	T. Harvey Pulling down old brick kilns and building new lodges	7	16	0
March 1	Hy. Barden Brick and Tile Making 1840	83	5	10
May 14	Geo. Geering Smiths work	2	11	7
	T. Dray Bricklayer to new brick kilns	12	0	4
	J. Baker Carpenters work	11	16	3
16	Brick Duty	1	10	7½
	Alfred Dawes Bricklayers work	8	2	6
		185	12	3½

Two kilns were built, each to take 20,000 bricks. Both were in existence 133 years later. However, by then their walls had become badly distorted, and only one of the kilns was considered safe enough for use during the last working years of the yard (a completed order in 1968 was a load of paving bricks for a car park in Storrington); they are primitive updraught kilns of the Scotch type, rectangular structures, open-topped, with a compartment for the fire underneath. Set in the side of an earth bank, each has an opening on one side for loading and unloading and on the other easy access to the fire tunnels in a roofed-over area.

In its first sixteen years the brickyard of 1840 made bricks almost entirely for works on the estate itself: only a few outside orders were undertaken. Edward Driver, a surveyor, had brought this situation about by saying in a report that too many of the farms had buildings of decaying timber, and that they should be replaced by brick buildings which would 'last for ever'.

The biggest single building ever served by the brickyard was Ashburnham Place itself. This had been stuccoed in the Regency manner during 1813, but by 1845 the stucco was cracking. The fourth Earl of Ashburnham decided to have no more rendered walls and set about completely re-casing his residence with Ashburnham bricks. The employment of grey headers in pink brickwork made a distinctive feature of the new walls. Expenditure recorded by the steward suggests that brick production was never greater than in the years between 1846 and 1855, when the house was being re-cased.

In 1847 the estate bought for £37 9s. an Ainsley Brick and Tile Machine, one of the earliest machines of the kind that were worth having. It appears to have been little used, though, and there is no indication that further items of machinery were bought later.

From 1856 onwards production was lower, but selling outside

the estate on a commercial basis began to take place. Commercial expansion between 1856 and 1869 clearly reflects the Victorian building drive which went on in many parts of England, not least in eastern Sussex. The area which the brickyard supplied in the second half of the nineteenth century extended from Burwash in the north to Hastings (east of Ashburnham) and to Eastbourne (west). For the same area *Kelly's Directories*, Mr Leslie discovered, show an increase in the number of commercial brickyards from eight to thirty-five.

The value of Ashburnham's outside trade, he points out, was that it encouraged the yard to stay in full working-order through times when there was little demand for bricks on the estate. Occasionally outside orders were embarrassingly large. In 1887 the Normanhurst estate at Catsfield of Thomas Brassey, the railway builder, ordered no less than 365,000 bricks and 36,500 plain tiles. Somehow these goods were mustered and dispatched, though a large number of the bricks were sent back on the ground that they were mossy, discoloured and chipped. It appears that to cope with the order, the foreman had included some very old stock.

From 1898 the yard sent few bricks into the outside world. It might well have been closed down in the 1920s, but Lady Catherine Ashburnham, who inherited the estate in 1924, favoured commercial brickmaking, and her interest accounts for the survival of the yard for a further forty years. The profits were never big: since the Second World War, the best was £316 for the year 1954.

Mr Leslie watched the yard in action during its last two seasons and has reported on the traditional procedures he saw. The brickearth, referred to as loam, was dug to a depth of about 7 feet—generally in the autumn only, to give the material a period of weathering by rain and frost. The day before brickmaking enough for one day's making was pulled down from a weathering heap with a mattock. After a day of being mixed up

85
Bricks being fired at Ashburnham in a kiln of the intermittent Scotch type. The photograph shows the entrance covered by a temporary brick wall; access to the wood-burning fire tunnel is the other side

with water, and pounded, it would be sufficiently plastic for moulding the following day (until 1961 Ashburnham had a horse-operated pugmill which relieved the men of much of the heavy labour of mixing).

Two moulders worked unaided in 8-foot square portable huts. Secured to the bench in each was a stock, hinged for easy removal of the moulded bricks. The mould, made of beech and shoed with iron, fitted over this stock and rested on four square-headed screws. Before each day's work it was necessary to check the height of these screws and set them according to the thickness of brick needed—sometimes thin paving bricks were ordered. The sand for dusting the mould was brought from a small quarry a quarter of a mile away.

The average production rate per moulder was between 500 and 600 a day, a low rate because both moulders had to perform all other brickyard operations. They had, for example, to wheel their green bricks to the hacks for drying, taking thirty-six in each barrow load.

The hacks consisted of concrete strips on which the green bricks were piled to a height of seven bricks; drying took between three and six weeks according to the weather. As soon as the bricks were firm enough to move about they were 'skintled'—that is, rearranged at an angle to catch the wind and draw it past them. The hacks ran north to south so that the bricks would get an equal share of sunshine on each side. As a guard against rain and frost, wooden hack-covers were placed on the heaps: in earlier days straw was invariably used for this purpose.

Mr Leslie found only one of the two Scotch kilns in use. When it was filled—a job which took three days—the loading hatch was sealed with a temporary brick wall to prevent loss of heat. The top of the kiln was covered with three layers of bricks laid sufficiently roughly to allow an updraught. Before the firing proper came the water smoking period, during which a steady fire drove off the water content remaining in the bricks. The fires were lit at the far end of the fire tunnels with the aid of paper and brushwood that blazed immediately and gave maximum draw.

After four days of water smoking at a temperature of about 240°C, the heat was gradually increased by pushing in more fuel in the form of underwood. To maintain the temperature needed, the wood had to be fed continuously to the fire tunnels, a job that, with ash-raking, fully occupied a burner throughout a shift of six hours. The burning process took about fifty hours, with the top temperature of 1,100°C reached after forty hours.

86
Newly moulded bricks being wheeled away for drying at Ashburnham

During the final ten hours the burner had to watch the colour of the fire tunnel walls; all was well if they remained orange-yellow or if—as had once been the saying—they matched the colour of a sovereign. It was also necessary to watch the dropping of the level of the bricks as they shrank under heat. When the degree of shrinkage had reached a certain point, the fire tunnels were not refuelled; their openings were closed over to encourage uniform burning by driving the heat to the top of the kiln away from the more well-cooked bricks at the bottom. After an hour the fires were charged again for a short period. Then the bricks were allowed gradually to cool. The whole process took five days.

About 75 per cent of the bricks taken out were first class in quality, a proportion still considered satisfactory for certain commercial kilns in use. They were sand-faced reds and not particularly attractive in colour; but in nearly all buildings in which the bricks were used the redness was reduced by ornamental placing of headers showing a pleasant blue-grey face. It is a peculiarity of wood-burning kilns, as said earlier, that the ends of bricks exposed directly to the fire turn out this colour: use of these for patterns in brick walls are a pleasant feature of much old brickwork in Sussex.

Britain's brick industry since 1900 16

Britain's brick industry is sensitive to quite small changes in the nation's housing programme, and every so often there is either a great surplus of bricks, as in 1969 when in some yards stacks awaiting disposal stretched as far as the eye could see, or a great shortage of the kind which became acute in early 1973.

87
The product. Six million bricks at Eastwoods' yard at Conyer, Kent—enough for 300 houses

At the works where bricks increase by the million every few days, lack of demand presents a worrying embarrassment. Victorian brickmakers were firm believers in the trade cycle and tried to be always prepared. Mr Morris Whitehouse, who is a member of a successful brickmaking family, has recalled that the seven fat and seven lean years were thought as certain as night and day. Mr Whitehouse's recollections were published in the *British Clayworker* in 1971, shortly after his retirement from the chairmanship of Redland Brick. He considered in them how far the theory has been borne out by the last seventy years.

The active years at the beginning of the century were indeed followed, he says, by an intense depression between 1905 and 1911 when there was great competition for custom among the brickmakers. The building trade as a whole was in a depressed state during that time. There had been rapid developments in the manufacture of Portland cement, with the result that builders were relying more and more on concrete. And steel, replacing the more expensive but splendid Victorian cast iron, was finding its way into many of the larger buildings.

The brickmakers responded with reduced prices for their bricks. They achieved these partly by faster and more economical production, but also by taking advantage of plentiful labour: although piece rates for workers became slightly lower than in the eighteenth century, if ever there was a

88
The Barnhart Steam Shovel of 1890; its cab could be securely locked at night

strike it was almost certainly short-lived and abortive.

A good result of the competition from concrete and steel was that the brick men developed multi-coloured facing bricks which replaced the Victorian even reds; the mellower bricks appealed to a new generation of architects, and encouraged such enterprises in public relations as the annual Building Exhibition.

Ironically, trade improved in the years just before the First World War, and 1913 was a prosperous year. It was prosperous, too, for the collieries, then privately owned, which had been turning the shale they dug up into bricks and now found that by modernising their plant they could add considerably to receipts from coal.

The year 1914 saw good trade. Few people thought that the war which began in the summer would last long or would much affect life in Britain. Even in 1915 it was business as usual. However, under the Defence of the Realm Act of the following

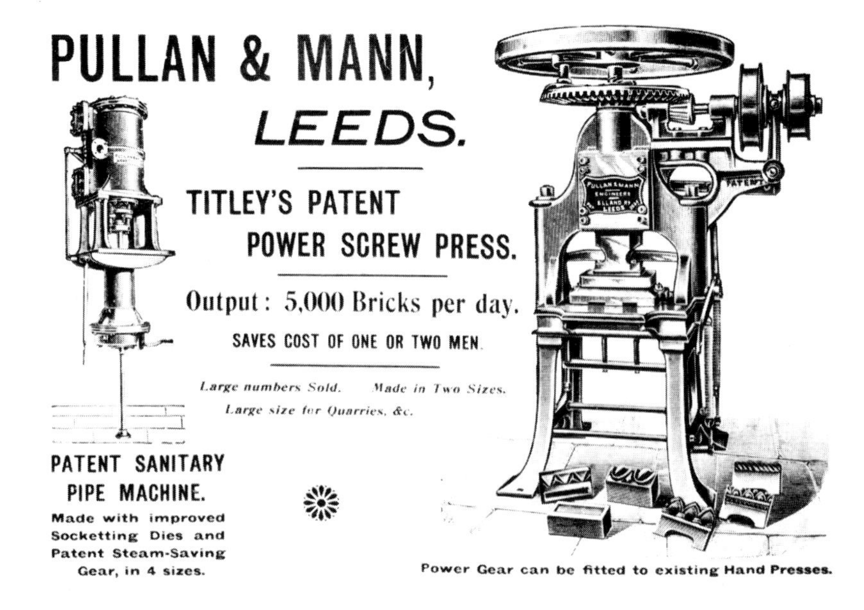

89
Advertisement in the *British Clayworker*, 1900

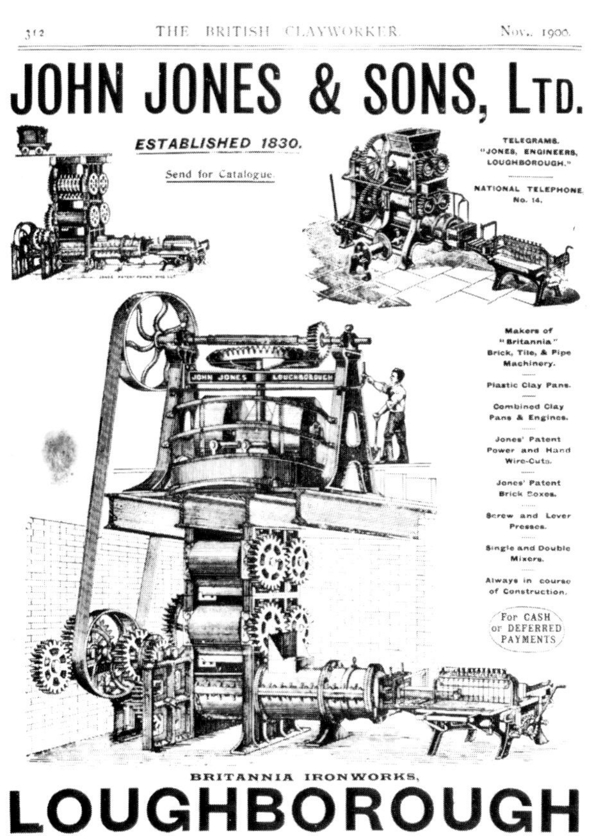

90
Advertisement in the *British Clayworker*, 1900

year, industry became mobilised for war and numerous brickmakers had to shut their works. Some recouped a little in being required to hire their kilns for the secret storage of explosives and ammunition; this was the excellent idea of J.W. Rowe, owner of the Star Brick Company at Peterborough and one of the pioneers of the fletton brick industry. His grandson is now Deputy Chairman of the London Brick Company.

91 *above*
Chambers' machine with an automatic wire cut-off in situ in America

After the war, large numbers of bricks were needed and 1919–22 was for the industry a period of recovery helped by the government's programme of housing, paid for out of public funds. But in 1922 it was decided that this cost too much and public building was cut back. The wheel seemed to be turning again. Brickmakers were suddenly in difficulties, not knowing what to do with all the bricks that mounted in their yards.

92
Brickmakers at work in the early nineteenth century

But in 1923 the operations of the speculative builder came to rescue them and the year 1924 saw a particularly heavy demand.

Denis Dighton
London. Printed & Published at Rowney & Forster's Lithographic Press 51 Rathbone Place. Feb.y 1821.

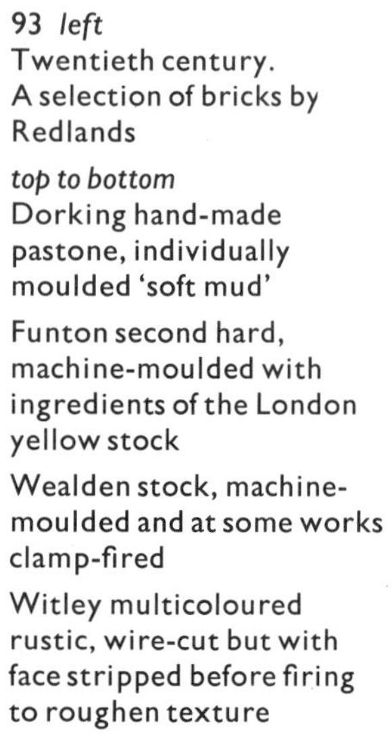

93 *left*
Twentieth century.
A selection of bricks by Redlands

top to bottom
Dorking hand-made pastone, individually moulded 'soft mud'

Funton second hard, machine-moulded with ingredients of the London yellow stock

Wealden stock, machine-moulded and at some works clamp-fired

Witley multicoloured rustic, wire-cut but with face stripped before firing to roughen texture

Soon builders were having such difficulty in getting delivery that a government embargo was imposed on the use of brick in public buildings. The shortage brought brickmakers the irritation experienced in recent times of learning that imports of bricks from Belgium had greatly increased.

The years between 1923 and 1929 were prosperous for the industry, if frustrating, and a time of investment and re-investment in brickmaking. Important innovations included better mechanical excavators, for which electricity was ousting steam as the driving power. In the transport of bricks there were changes of consequence in that reliance on the railways became less universal; increasingly steam lorries and petrol-driven lorries drove loads direct from works to building site.

The opportunities for road haulage were of great benefit to small country brickyards which hitherto had been limited to their local market. It can be added that the quality of their bricks was often so high that the image of the brick industry as a whole was improved by their increased sales.

Rail transport was however still cheap. Some of the railway companies were having a bad time and would quote low charges to brickmakers to maintain the business; but the problems to do with delays and damage, and the expense of unloading by hand into road vehicles at the railhead, were never really solved. Road transport gradually took over the majority of brick traffic.

After another period of lean years following the slump of 1929, the brick firms once more found a ready market in 1933 when the government instituted, largely to provide work for the unemployed, a programme of public building work. By 1934 the speculative builder was again busy and the next four years were prosperous for brickmakers.

Reduced building activity in 1939 meant that by the outbreak of the Second World War brickmakers had accumulated large

94
The washback. A means, now in the past, of preparing the material for bricks of London-stock type: a slurry of clay and chalk was pumped into a hollow and allowed to settle. The photograph shows one of Eastwoods' washbacks in the 1940s

95
Women working on brick presses in the fletton brick industry during the Second World War

stocks. Kiln fires then became a landmark at night for enemy aircraft and, as arranged by the Simmonds Committee, numerous works closed for the war years, and once more kilns were used for storing ammunition. But the closures were not carried out in the haphazard manner of the First World War—after which many plants never opened again.

The leaders of the industry called to serve on the Simmonds Committee had worked out an orderly plan for closure. There was a levy on firms remaining in business out of which a fund was accumulated for the purpose of putting closed firms on 'care and maintenance'; and a further fund was set aside for helping firms to re-open after the war. These far-sighted moves kept the industry in being and ensured that the immense quantity of bricks needed to repair war damage was available when the time came. There had never been a more satisfactory arrangement for tiding over a particularly lean period.

96
Stacking bricks of the former Dorking Brick Company in a kiln, 1946. The pile of battered bricks outside were used again and again for making the temporary wall to seal the kiln

The period 1949–54 was a good one, sales rising from 4,500 million to just over 7,000 million. These years saw the introduction of special fork lift trucks for moving suitably formed stacks of bricks, reductions in labour costs through mechanisation generally, appreciation of quality control, increased support for the British Ceramic Association and the setting up by leading firms of their own laboratories. In this period the industry became alert for the first time to the need to reorganise itself with a view to dealing with the recurrent problems of recession.

A process of amalgamation and take-over set in soon after the war, and an estimated total of 1,316 separate brick works in 1950 had dropped by 1967 to 581; however, the capacity for producing bricks was greater by about 1,000 million (in Mr Whitehouse's view) than the capacity of 9,300 million in 1950.

While the brick firms rationalised themselves into larger groups, London Brick turned itself into a very large undertaking indeed, and showed the world that its cheap flettons could compete effectively with all other building materials. By the beginning of 1973 it was making just over 43 per cent of all Britain's bricks. Another fifteen concerns in that year shared the making of the next 35 per cent of the bricks, while the rest of the industry's production was scattered among small producers—to give a total of perhaps 300 brick firms.

Any consideration of the future of the brick industry in Britain brings up the question of how much we need to go on using bricks rather than other units which, through being larger, are less expensive to lay. After the Second World War much taxpayers' money was devoted in the ministries and at government research stations to the development of new systems of building which went beyond mere changes in the size of bricks. Official encouragement was given to local authorities to try them out, with the result that new methods of building proliferated: only a proportion of these made any use of brick. Local authority approvals for dwellings put up in the industrialised manner rose from 30,000 (21 per cent) in 1964 to 71,000 (42 per cent) in 1967; but there was a reduction in 1968 to 60,000 (39 per cent).

97
Laying LBC 'through-the-wall' blocks, grooved for a rendered finish

The subsequent decline in the use of industrialised building—following a serious credit squeeze and the Ronan Point disaster in which the slab construction of a London tower block proved all too vulnerable—reinforced the view held by leaders in the brick industry that the arguments in favour of traditional building had much to commend them.

The use of concrete blocks in internal walls had accelerated greatly after 1945, to the detriment of common brick sales. Although the brickmakers were aware of this development, the speed with which it was happening was masked by the fact that, because of intensive building in the immediate post-war years, the number of common bricks sold was continually rising. Nevertheless the percentage of common bricks in their total sales fell from 82 per cent in 1946 to 70 per cent in 1955. There has been a further decline in the sale of common bricks since then, though it appears to have come to a halt more recently.

Persistence of demand for common as well as facing bricks is a good omen for Britain's brick industry. There seems, indeed, every reason why the industry should survive, throughout lean years and fat. The geology of the country provides the raw material for economic and varied production; brick suits a climate calling for a material that resists rain, sun, humidity and air pollution; it is cheap to buy, flexible to build with and costs nothing at all to maintain.

Brick is associated with a pleasant tradition in the United States, too, despite the taste for wooden houses. America's brick industry reached a peak in production just before and just after the First World War—the record figure was 15½ million bricks in 1909. There was a big decline after 1925, when mass brickwork no longer made skyscrapers, and by 1950 annual production was down to 7,500 million, a smaller number than for

98 *left*
Grinding the clay at a modern fletton brick factory

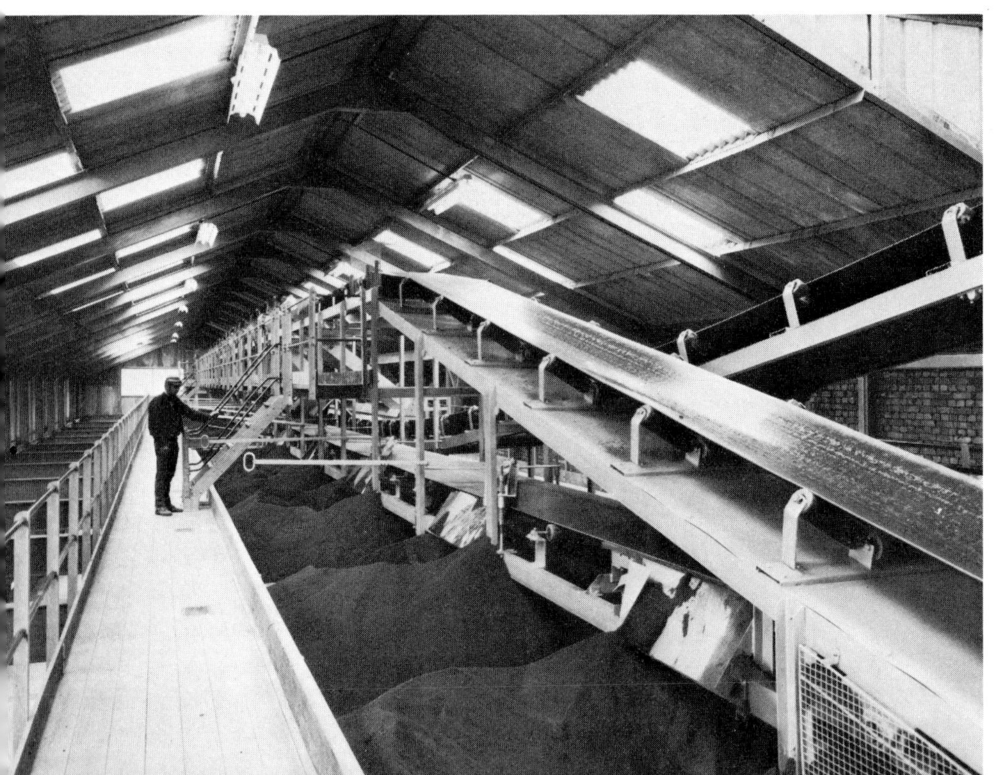

99 *left*
Screened clay being fed into the hoppers above the brick presses at a modern fletton brick factory

100 *right*
Unburnt pressed bricks being conveyed from brick presses at a modern fletton brick factory

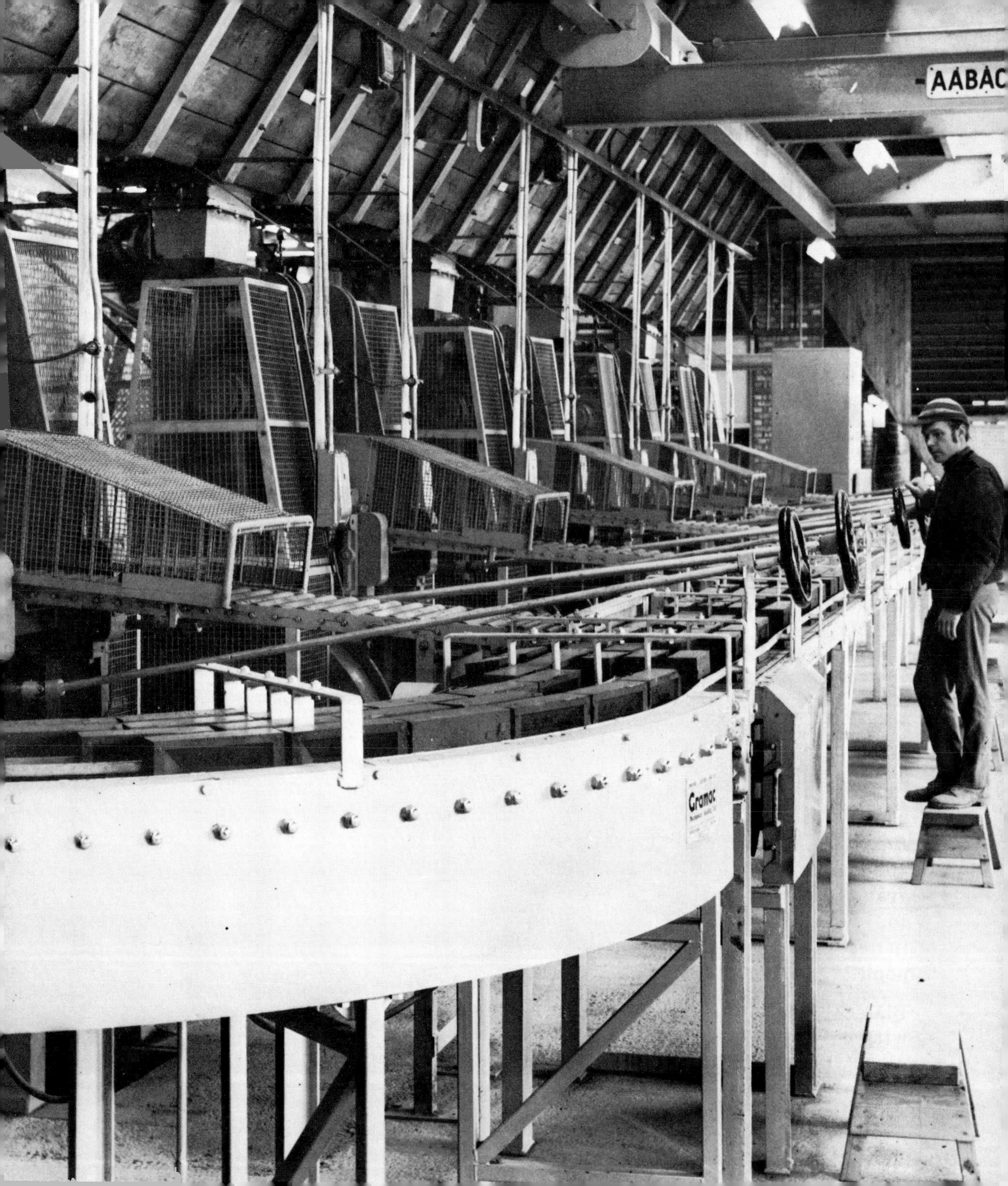
AABAC
Gramac

England in 1970. Bricks were being overtaken by large blocks of breeze or cement which were cheaper to lay.

But although in the States common bricks have been virtually unsaleable since the mid-1950s, such is the prestige of brick that people are prepared to pay well for the luxury and appeal of a brick finish, even though the price at the end of perhaps a thousand mile journey is often excessively high. The sales of brick in the United States have, indeed, picked up slightly in recent years and were in 1974 equivalent to those in Britain. That this is so with the United States having almost four times the population of Britain, and forty times the acreage, plainly reflects the different patterns in use of building materials in the two countries.

Britain's Brick Development Association, which exists to promote the use of bricks in this country, emerged in 1954 from the foundations of the wartime National Brick Advisory Council. BDA now represents the whole of the brick industry including the manufacturers of sand-lime bricks. The association's financial strength dates from 1964 when government funds ran out and members were required to pay a subscription based on their turnover. In 1973 the BDA had an annual income of nearly £500,000 with which, from its London headquarters, it seeks to supply information to the building trade and generally to advance the cause of brick.

Information sheets and pamphlets are constantly sent out to people concerned with building. A colour journal called the *Brick Bulletin*, showing up-to-date uses of brick, goes free to about 20,000 people in most parts of the world. Excellent photographs show the brickwork ideas of one country to another; so, even if huge pre-fabricated panels of brickwork are little in demand in Britain, at least it is widely known that they are popular in parts of northern Europe.

BDA acts today as a mouthpiece for the industry in its relations with government and as a co-ordinating body on matters

101
A prefabricated brick panel complete with cavity

affecting research and education. In a competitive industry, it seeks to provide a forum where brickmakers large and small can discuss matters of mutual interest.

Brick tiles 17

Clay roofing tiles are made in much the same way as bricks. But because roofing tiles are larger than bricks, and yet only half an inch thick, the clay must always be mixed to a very fine texture to avoid distortion during the burning. Tiles are moulded, or else wire-cut—that is, cut off from a column of paste forced through a die. The nail holes are formed afterwards with a punch.

The plain flat tiles so common in Britain appear to have come into general use before bricks. Mr Clifton-Taylor considers it was not unusual by the year 1300 for the better kind of house in the east and south-east to have a tiled roof—and he draws attention to the incidence in these parts of the surname Tyler.

102
Plain tiles being moulded by hand. The procedure is as for bricks, only the stock is raised higher to give the right thickness

References to mediaeval tilemaking west of the limestone belt are few, but in the 1830s Staffordshire and Shropshire, helped by canal transport, became important tilemaking counties. The north of Britain had few tiles; stone and slate were available for roofing and these materials harmonised with stone walls.

Plain tiles began as copies in clay of shingles. At first there was a confusing multiplicity of sizes in use, but the size laid down in the reign of Edward IV, $10\frac{1}{2}$ inches by $6\frac{1}{2}$ by $\frac{1}{2}$, proved so convenient it has not been changed in five hundred years. The curved pantile, of Dutch origin, was introduced to England in the seventeenth century and is common on roofs in East Anglia and the north-east. Until recently the making of tiles and bricks often went on in the same works, and both have been subject to great improvements in machine production.

Clay for tiles, whether plain or of an interlocking type, has today been largely overtaken by concrete. However, so many

103
Mathematical tiles, header and stretcher

104
Mathematical tiles, resembling bricks, were sometimes used in conjunction with real bricks, as at No. 33 High Street, Lewes, where the existence of tiles on the side wall is betrayed by certain depressions and lines. Angles could be difficult with mathematical tiles: here large corner stones simulated in wood cover the quoin joints

people prefer the clay kinds, which weather even more pleasantly than bricks, that a minor revival in their manufacture, making use of the modern tunnel kiln, could well take place. It is also possible, if only for the sake of keeping certain eighteenth- and nineteenth-century buildings in repair, that once again the mathematical tile will become available new as well as second-hand. Since this particular tile has no function other than to cover walls, it can be considered in a book on bricks even more readily than roofing tiles.

Mathematical or brick-tiles are tiles so shaped that when fixed to a wall and bedded round with mortar, they resemble brickwork. They are seen for what they are only when there is imperfect work on window surrounds or vertical corners, though of course the game is given away by a nail failure which has let a tile start to slip out. The main function of these tiles was to make wooden houses appear to be brick. In towns especially, the Georgians preferred brick to wood, and caused a profusion of houses to be given the required look by means of brick-tiles in the south east towns of Lewes, Rye, Tenterden and Brighton.

The tiles have often been convincing enough to deceive architects and demolition contractors—not to mention owners of houses covered with them—but they were an efficient waterproof cladding as well as a deception, and were often referred to as weather-tiles in the eighteenth century; indeed it has been claimed that mathematical tiles offer one of the best means ever devised of shedding water from a vertical surface. Normally they fail only when there is movement in a timber framing.

Who invented them is not known—at any rate, not yet. The term, mathematical (geometrical has also been used), was presumably suggested by the regular and precise face of these tiles compared with overlapping plain tiles, which in the eighteenth century were thought rustic and inelegant.

It has often been said that mathematical tiles were invented to because they appeared before its introduction. Nathaniel Lloyd records their use in 1755 at Lamb House, Rye, Sussex, and John Archibald in *Kentish Architecture Influenced by Geology*, 1934, says they were used from 1725.

There is no doubt, though, that the brick tax was an incentive for putting them on new buildings. When in 1803 roofing tiles as well as bricks had to bear a tax, it was found that the mathematical tile, a mixture of the two, was not subject to it. But mathematical tiles continued to be made for some years after the brick tax was repealed in 1850. They were recommended in successive editions of J.C. Loudon's mid-nineteenth-century *Encyclopaedia of Cottage, Farm and Villa Architecture*:

> The walls of cottages may be protected and ornamented by mathematical tiling. The object of this is to make the walls appear as if they were built of brick. The tiles have their surfaces in two planes. . . . There are bats or headers to imitate half bricks, and closers or quarter bricks for the purpose of breaking joint at the angles and making the imitation more complete. When these tiles are of cream colour, their effect is very neat, clean and handsome . . . .

The tiles were being made again by the Keymer Brick Company of Burgess Hill during the 1960s and used for a few new buildings as well as for repairing old ones. Keymer called their tiles Keytoclad tiles and pointed out in the brochure that they inherited all the attributes of clay: its warmth, mellowness and dignity. 'Keytoclad tiles have a multplicity of uses ranging from the facing of a block of flats to cladding a timber garage. Keytocladding reduces skilled man-hours in building facing work without in any way detracting from the appearance desired.'

In the Sussex Archaeological Museum at Lewes, there are examples of the early Lewes tiles, made in strips with $4\frac{1}{2}$-inch

105
Twentieth century. A selection of London Brick products which have either, been scored and dented before firing, or received blasting treatment with grit. Many experiments have gone into finding out how to produce a finish that people like

above left to right
Rustic, invented 1922; Tudor, 1962

centre
Golden Buff, 1954; Milton Buff

below
Cotswold, 1967; Dapple Light, an effect achieved with powdered steel slag

106 *left*
Panels prepared with two of the bricks in Plate 105 London Brick's Golden Buff and Tudor—which show the effect of changing the mortar colour

shallow vertical incisions to facilitate splitting into either headers or stretchers. Provision was also made at the end of each strip for splitting the $4\frac{1}{2}$-inch header into two $2\frac{1}{4}$-inch closers. Blue tiles were always made as separate headers to avoid accidental exposure of the red colour under the blue film.

Mathematical tiles are to be found in a few places outside the south-east. There are some in Durham, Horncastle and Cambridge and at Attingham Park in Shropshire. Mr C.G. Dobson was surprised to find them in Wandsworth, London. In March 1965 he wrote two helpful articles on what he called the gay little deceivers in the *Illustrated Carpenter and Builder*, and described mathematical tiles that were the facing material of the ground and first floor of a house at 14 Garratt Lane, Wandsworth. They had become insecure and the walls were covered with vertical bands of stout hoop iron to keep them in place. Mr Dobson believed that the reason for the failure to adhere properly was the weight of a further storey, a late addition, which had caused some distortion of the lower timber framing.

107 *below*
Directions for hanging mathematical tiles in a brochure of 1964 issued by the Maidenhead Brick and Tile Company. Regular production of the firm's mathematical tiles, which they called Keytoclad tiles, ceased in 1965

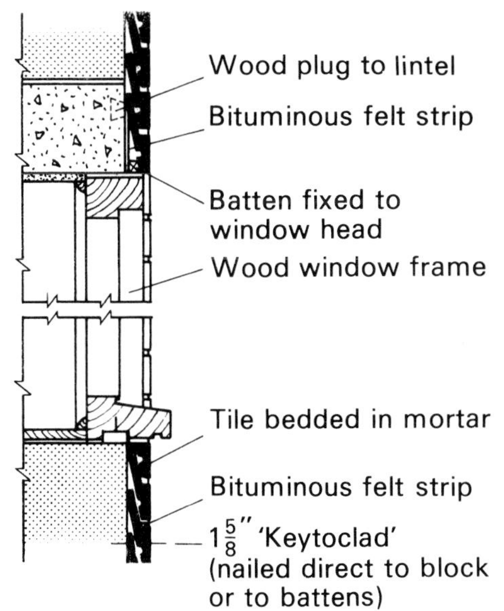

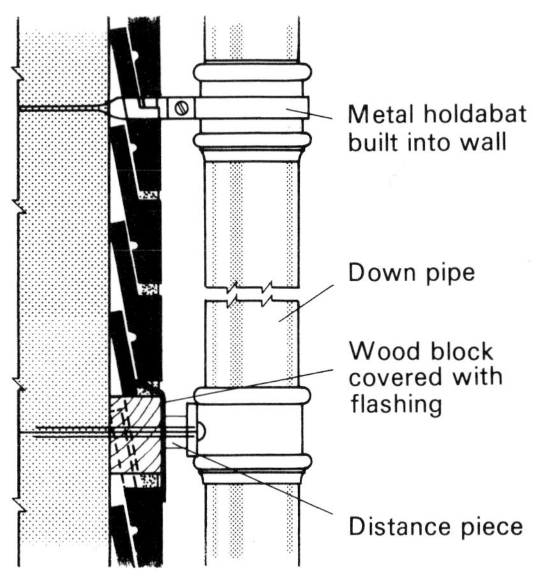

The tiles were by no means only for timber-framed buildings. There are examples of their use on existing brick walls to change the colour (usually from red to yellow) and as part of an elaborate redesigning to keep up with current fashion. A large number of brick houses of Stuart type were given a classical façade in Georgian times by changing the windows and fixing mathematical tiles all over the earlier brickwork. New brick houses were sometimes given the treatment, too. The Royal Crescent in Brighton, dated between 1799 and 1807, is a terrace of tall red-brick houses which were faced all over with black mathematical tiles. They cost little to maintain compared with Brighton's stuccoed buildings.

Stonework, too was occasionally masked in this way—perhaps to fit in with a brick addition. When electric light was being installed at Nunwell House near Brading, Isle of Wight, the people living there were much surprised to find that their entrance front, apparently brick, was in fact of hard stone. A skin of red mathematical tiles had been skilfully applied.

18 The heritage of brick

In times of rapid change people like to look again at the enduring legacies of the past. Those of brick can be worth looking at. Building for building, they have weathered better than the two heritages forced upon us by wartime necessity and experience, 'the pre-fab (leading to Ronan Point) and shuttered concrete (leading to the Hayward Gallery)'. The phrase is Alistair Horne's. Time, he has written, has been kind to various examples of wartime concrete but not so kind to the Hayward Gallery on London's South Bank, 'which seems already to be ageing less gracefully as the stains take hold than the bunkers of France's Maginot Line'. Concrete is widely used today, often doing in a seemly manner a job which no other material would do; but can it be said that in the public esteem it has had its finest hour—in the Second World War?

Some brick is as ugly as some concrete (stone, too, can be repellent), but no building material except brick offers so wide a range of colour, so much scope for invention, such permanent economy of maintenance. Although few (wasteful) solid bricks are being made today in Mediterranean lands, there is for visual purposes an increasing employment in conjunction with whitened stucco, of the smoothly pink 10-inch brick, light in weight because of 50 per cent vertical perforation. Many a new Italian villa has sunlit pastel-colour brickwork for its ground floor storey and pastel-colour pillars at its gates. In a mountain village near Carrara called Fabbiona, where even cottage roofs, pathways and garden walls are made of marble, the inhabitants recently accepted pink bricks for the surround of their communal water pump in the tight village square.

Britain's heritage of brick is palpable in Tudor manor houses, classical Georgian mansions and in some of the magnificent buildings of the first half of the so-called Industrial Revolution of roughly 1775 to 1840. Consider the huge mills of the north and the first railway undertakings. They have the majesty of proportion associated with good stone masonry. The series of Port of London docks, built in the first decade of the nineteenth century, forms a tremendous range of brick construction which is probably unrivalled. Brunel's bridge over the Thames at Maidenhead, 1838, has the largest arches, and the flattest in proportion to their span, ever made in brick. They are semi-elliptical, each of 128 feet span with a rise of 24 feet—L.T.C. Rolt, in *Isambard Kingdom Brunel*, 1957, has graphically described how the eastern arch had to be rebuilt because the centerings were removed before the Roman cement had properly set and some distortion took place. The early waterworks like those at Hammersmith and Pimlico are good specimens of London's early industrial buildings done in brick (the Pimlico one has recently been cleaned to its original yellow). All over Britain there are plenty more that are as well built, well suited for their jobs and pleasing to the eye.

Numerous over-elaborate brick buildings appeared in the second half of the nineteenth century, a result of the movement for enlightened architecture begun by the Great Exhibition of 1851: carpet factories modelled on the Alhambra, warehouses suggesting Gothic cathedrals and red office buildings, like the Prudential in Holborn, worked up in the grandest picturesque manner. The showiness of Birmingham University, with its Sienese tower by Sir Aston Webb, helped to establish for the newer non-Oxbridge universities their unflattering label of red-brick.

Florid Victorian buildings in red brick have nevertheless survived barrages of criticism. And these criticisms are not solely of twentieth-century, post-Lutyens origin, for Victorian brickwork had Victorian detractors. Mrs E.T. Cook, writing *Highways and Byways in London* at the end of the nineteenth

108
A house by Sir Edwin Lutyens presenting his own version of English garden wall bond (more header courses than usual)—Middlefield, Shelford, Cambridge, 1908. The brickwork of the chimneys is only at first sight identical: in the middle chimney the centre is recessed while in the other two the centre projects, a variation which neatly avoids unresolved trinity. The roof of the house has been described by Christopher Hussey as 'one of the loveliest, surely, ever raised, set continuous over every part, its valleys trimly swept . . .'

century, was only one of many to censure it. She deplored the new hotels of 'ornate red brick' and the 'enormous Mansions . . . hardly attractive' which had risen in Babel-like height. 'Everywhere', she wrote, 'is red brick and red or buff terracotta, adorning alike shop front, warehouse, tube station and palatial mansion.'

109
Brick architecture in Lutyens's Romantic manner—the Deanery Garden, Sonning. Christopher Hussey: 'a perfect architectural sonnet, compounded of brick and tile and timber forms in which the handling of the masses and spaces serves a rhythm'. The vaulting here is of hewn chalk

In the domestic sphere, great pains would be taken by an architect and his client to make sure that brick houses had an ornamental appearance. In Wimbledon, Hampstead, Hove and Bournemouth—wherever there was money to spend on such things—there are to be seen houses of rather shiny red bricks relieved with salmon-coloured terra-cotta work and griffin terminals to the ridge tiles. To their first owners they were in the tradition of the English gentleman's house. Today we have so many buildings over which no trouble seems to have been taken to please the eye that we are ready to look admiringly at Victorian brickwork. All the same, it can take a slight mental effort to do so.

In the entirely easy exercise of admiring pre-Victorian brick buildings, what is to be noted is that the actual bricks as well as the designs are different: they are not machine-made, they are not uniform in shape and colour or shiny. Up to about 1840 the bricks employed for churches, factories and most engineering work were of the same hand-moulded kind that went into houses and cottages. The effect of the individual bricks is such that it still governs our response to the heritage of brick buildings: for this reason an architect will often turn to a small hand-making yard when bricks of gentle appearance are needed to repair or add to an old house.

Although machine-made bricks have long dominated the market, some facing bricks are still hand-made as a matter of course. People like them for the crease marks and for the minor irregularities of shape and colour, and they are prepared to

pay three times the machine-made price. This is not so extravagant as it seems, because the level of costs in building a house is now such that the price of the bricks may not be more than a small fraction of the bill.

The number of hand-making works has actually gone up in the last few years, for closed ones have opened again. The output of hand-made bricks in 1973 was about 2 per cent of the total—perhaps 150 million a year. There is, too, greater reliance on the old clamp firing method (over 200 million a year never go in a kiln), because of the varied colours produced.

In hand-moulding, a clot of clay must be so thrown down that it entirely fills the sanded mould; and dexterity is needed with the pallet boards to transfer the soft brick to a wheeled rack. To train a moulder may take two months. It is heavy work, the incentive of piece rates leading to great exertion, and many a beginner has badly strained his wrists. Today's moulder tends to work on his own instead of as one of a gang; but having to break off to wheel his bricks away to the drying shed is believed to be beneficial in giving him respites from moulding. Single brick moulds are usual, though seasoned moulders at Colliers, Marks Tey, Essex, handle triple-brick moulds.

110
Hand-moulding bricks at Bulmer's Suffolk yard in 1972. The brickmaker's name, significantly, is of Dutch origin—Peter Kloos

Calling at a Buckinghamshire brickyard just before closing, I found a couple of young moulders working as though in a race, yet never failing to lift the clay above their heads before dashing it into the moulds. The foreman confirmed that they had been going at that pace since six o'clock in the morning. 'How many today?' one was asked. 'About 1,650—more than yesterday.' And as he paused to speak, his arms and hands went on moulding bricks in mime. Brickmaking is one of a very small number of manufactures whose ancient methods are still to an extent in profitable use, providing a living heritage.

But in considering brick's heritage in the more conventional sense of the word, it can be said that brick has bequeathed to Britain a legacy on the land in addition to the buildings: the sites so often indicated by such place names as Brick Kiln Farm, Kiln Wood, Brickfield Cottages and Claypit Lane. Who has not come across one of them? The traces of defunct brick-yards are scattered over the landscape.

Mr W. Howard Williams has personally tracked down over seventy sites in Shropshire and Mr John R. Jackson, a British Brick Society member, has found about twenty in the southern part of Essex. At least two museums are actively interested in the heritage of brickmaking. A few years ago the Brideswell Museum of Local Industries, Norwich, undertook to collect at least one brick from each old kiln site in Norfolk—and here and there, moulds. Where kilns and sheds still exist, these are being photographed; former workers are interviewed and tape-recorded. At Little Plumstead, where ornamental bricks used to be made, eighty-three examples, thirty-four wooden moulds, two barrows and some hand tools have been acquired. Workers for the museum—many are school boys and girls—also collect

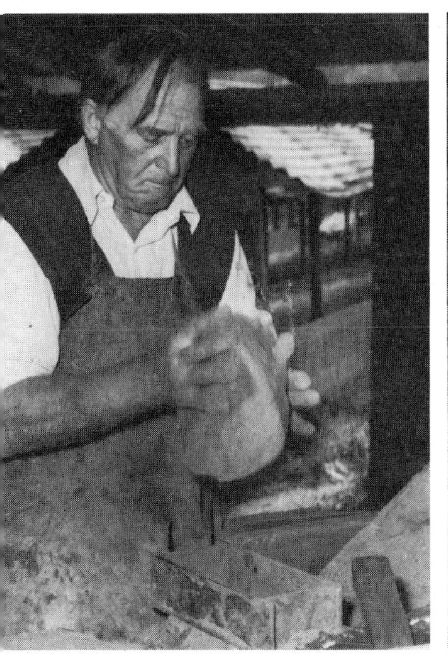

bricks from old buildings where these are being pulled down, or where advice about alterations has been asked for. Exactly similar work is being done by the Doncaster Museum.

The heritage of brick that consists of excavated stumps of mud brick walls built thousands of years before Christ, as described in chapter 3, has now become pointedly related to present-day needs abroad. Indeed the sun-dried mud brick is being spoken of as the best answer to the world's housing shortage of around 760 million dwellings (a United Nations figure), since there are huge areas of the world where fired bricks and concrete are out of reach.

Over half the world's population is said to live in mud houses which were often built at the cost of the occupant's labour only. But ordinary mud bricks are soon damaged by water and walls need re-plastering with a mud mortar after each rain. One of the best ways so far known of internally waterproofing mud bricks

111 *left*
Moulding mud bricks in East Africa

112 *right*
Straw being included as a binding agent in Mexican mud bricks. Straw, as in mediaeval times in Britain, also makes the platform for the drying bricks, but the shape of these bricks is ingeniously modern and designed to tie together the two leafs of a wall

is to add to the mix about 2 per cent by weight of asphalt emulsion. This has been done in the southern states of America since the 1930s—even for rich men's houses—and to a small extent since the mid-1960s in Iran. The Shah of Iran became interested when he watched a demonstration in which the Teheran Fire Brigade played water on some walls of asphalt-stabilised bricks and saw that, unlike adjacent walls of ordinary mud bricks, they could not be washed away.

Plenty of countries have no ready access to bituminous products, however, and in 1974–5 British scientists at the Building Research Station near London were taking part in international efforts to find new ways of using lime to stabilise earth—and economical ways of producing lime on site. Whatever the results may be, it is interesting that the sun-dried mud brick, with which this book began, should remain demonstrably a vital building material.

Select bibliography

Banditt, W. O., *Gebrannte Erde*, Steinbock Verlag, Hanover, 1965.

Bardwell, W. *Healthy Homes and How to Make Them*, Gilbert, 1860

Barley, M. W., *The English Farmhouse and Cottage*, Routledge, 1961

Bourry, E., *Treatise on Ceramic Industries*, Scott Greenwood, 1901.

Braun, H., *Old English Houses*, Faber, 1962.

Butterworth, B., *The Properties of Clay Building Materials*, British Ceramic Society, 1953.

Butterworth, B., and Foster, D., *The Development of Fired-Earth Brick*, British Ceramic Society, 1958.

Clew, K., *The Kennet and Avon Canal*, David & Charles, 1958.

Clifton-Taylor, A., *The Pattern of English Building*, Faber, 1972.

Davey, N., *A History of Building Materials*, Phoenix, 1961.

Davis, C. T., *Treatise on the Manufacture of Bricks, Tiles and Terracotta*, Philadelphia, 1895.

Dobson, E., *Treatise on the Manufacture of Bricks and Tiles*, London, 1850.

Every, G., *Christian Mythology*, Hamlyn, 1970.

Fletcher, V., *Chimney Pots and Stacks*, Centaur Press, 1968.

Forrester, H., *The Smaller Queen Anne and Georgian House*, Tindal Press, 1964

Harley, L. S., *Polstead Church and Parish*, British Publishing Company, Gloucester, 1965.

Hartley, L. P., *The Brickfield*, Chatto & Windus, 1964.

Hudson, K., *Building Materials*, Longman, 1972.

Hussey, C., *The Life of Sir Edwin Lutyens*, Country Life, 1953.

Kenyon, K., *Digging Up Jericho*, Benn, 1957.

Laurie, A. P., *Building Materials*, Oliver & Boyd, 1922.

Leslie, K., 'The Ashburnham Estate Brickworks', *Sussex Industrial History*, 1971.

Lloyd, N., *A History of English Brickwork*, Montgomery, 1925.

Loudon, J. C., *Encyclopaedia of Cottage, Farm and Villa Architecture*, 1833.

Malet, H., *The Canal Duke*, David & Charles, 1961.

Marshall, S., *Fenland Chronicle*, Cambridge University Press, 1967.

Mellart, J., *Catal Hüyük*, Thames & Hudson, 1967.

Rolt, L., *Isambard Kingdom Brunel*, Longmans, 1957.

Ross Williamson, R. P., 'The Progress of Brick Through English History', *Architectural Review*, May 1936.

Salzman, L. F., *Building in England*, Oxford University Press, 1952.

Searle, A. B., *Clays*, Pitman, 1915.
Modern Brickmaking, Benn, 1956.

Seger, H. A., *Collected Writings on the Manufacture of Pottery, Vol. II*, Scott Greenwood, 1902.

Smith, G., *The Cry of the Children from the Brickyards of England*, 1871.

Stratton, P. M., 'The Bonding of Brickwork', *Architectural Review*, May 1936.

Vitruvius, M., *The Architecture of Marcus Vitruvius Pollio*, translated by J. Gwilt, Lockwood, 1874.

Wight, J., *Brick Building in England from the Middle Ages to 1550*, Baker, 1972.

Willcocks, W., *From the Garden of Eden to the Crossing of the Jordan*, Spon, 1920.

Williams-Ellis, C., *Building in Cob, Pisé and Stabilised Earth*, Country Life, 1947.

Willmott, F. G., *Bricks and Brickies*, Willmott, 1972.

Woodforde, J., *The Truth about Cottages*, Routledge, 1969.

Index